Total Organizational Excellence
Achieving world-class performance

Professor John S. Oakland
Executive Chairman
Oakland Consulting plc

Professor of Business Excellence
Leeds University Business School

BUTTERWORTH
HEINEMANN

AMSTERDAM BOSTON HEIDELBERG LONDON NEW YORK OXFORD
PARIS SAN DIEGO SAN FRANCISCO SINGAPORE SYDNEY TOKYO

Butterworth-Heinemann
An imprint of Elsevier
Linacre House, Jordan Hill, Oxford OX2 8DP
200 Wheeler Road, Burlington, MA 01803

First published 1999
Revised paperback edition 2001
Reprinted 2002, 2003

British Library Cataloguing in Publication Data
A catalogue record for this book is available from the British Library

ISBN 0 7506 5271 3

For information on all Butterworth-Heinemann publications
visit our web site at www.bh.com

Typeset by Avocet Typeset, Brill, Aylesbury, Bucks
Printed and bound in Great Britain by Biddles Ltd, *www.biddles.co.uk*

Total Organizational Excellence

To the memory of Anne, my mother
and John Glover, my friend and colleague.

Contents _____

Preface _____

I wanted to call this book *From Head to TOE* (Total Organizational Excellence), but my editor said, 'It sounds like a book on fitness.' My reply was not profound, 'It is – it's a book on organizational fitness.' The 'head' is, of course, the top management in the organization – the chief executive and his/her colleagues who are responsible for the performance of the company, enterprise or public sector organization. Organizational excellence must spread from the top right through the body to the toe.

In the book, I have set down a framework or blue-print for achieving world-class performance. Based on many years of research and advisory work in Oakland Consulting plc and its research and education division, the European Centre for Business Excellence, the book guides senior managers through the framework, which has already achieved wide acclaim when presented at conferences and seminars throughout the world.

The first version of the framework was published in *Cases in Total Quality Management* which I co-authored with my colleague Dr Les Porter. The model appeared in the case study on goal deployment in Exxon Chemical, from the work largely led by Alan Randall. A development of these initial ideas then appeared in another book from the Oakland Consulting stable, *Assessing Business Excellence*, by Les Porter and another colleague Dr Steve Tanner.

The first chapter describes briefly the framework and each subsequent chapter examines a part in detail, which is highlighted on the blue-print diagram at the beginning of each chapter where appropriate.

The book has been written primarily for practising executives and managers in both the private and public sectors, commercial and non-commercial. Illustrations from real organizations are a feature of the book, which I have tried to write in a down-to-earth, practical style. The book should also be suitable for students in business schools, particularly at post-graduate level.

I could not have written the book without the tremendous energy and support of my close colleagues in Oakland Consulting. Several people have directly contributed to the development of the framework and some of the detail in the chapters. In particular I should like to mention Roy Broadhouse, Danny Burke, Ken Gadd, Nigel Kippax, Susan Oakland, Les Porter and Steve Tanner. Julie Wilson and Janet Selby have helped in keeping me sane whilst getting the detail together.

John S. Oakland

Chapter 1

A framework for total organizational excellence _____

Key points

This introductory chapter shows how organizational excellence may be integrated into the strategy of any business through an understanding of the core business processes and involvement of the people. This leads through process analysis, self-assessment and benchmarking, to identifying the improvement opportunities for the organization, including people development.

The identified processes should be prioritized into those that require continuous improvement, those which require re-engineering or redesign, and those which lead to a complete re-think or visioning of the business.

Performance-based measurement of all processes and people development activities is necessary to determine progress so that the vision, goals, mission, and critical success factors may be examined and reconstituted if necessary to meet new requirements for the organization and its customers, internal and external. This forms the basis of a new implementation framework for total organizational excellence which provides the structure of the book.

Avoiding the confusion

'Total quality management (TQM) is dead, long live business process re-engineering (BPR)!' 'ISO9000 is too costly/narrow focused, you should carry out self-assessment to the European Excellence or Baldrige Quality Award models.' 'Statistical process control (SPC), failure mode and effect analysis (FMEA) and benchmarking – these are things you should be using.' And what about measurement, culture change, teamwork, continuous improvement, etc., etc.? My goodness, no wonder people are confused and irritated by the conflicting messages (and combination of letters) they now receive from consultants, academics, business leaders and even politicians about what they should do to improve the performance of their organization.

To try to get some of these things into sensible proportions and shape, my colleagues and I in Oakland Consulting have carried out a great deal of research, teaching and advisory work. The people we speak to often are confused and in urgent need of a framework to pull this lot together, and I am sure this is no different in your organization. This book attempts to provide that framework – a blueprint for total organizational excellence.

Firstly, let me say that TQM is not dead and is not the very narrow set of tools and techniques often associated with failed 'programmes' in organizations in various parts of the world. It is part of a broad based approach used by world-class companies, such as Hewlett Packard, Milliken, TNT, and Yellow Pages, to achieve organizational excellence, based on customer results, the highest weighted category of all the quality and excellence awards. Total organizational excellence embraces **all** of the areas I have mentioned so far. If used properly, and fully integrated into the business, this approach will help any organization deliver its goals, targets and strategy, including those in the public sector. This is because it is about people and their identifying, understanding, managing and improving processes – the things any organization has to do particularly well to achieve its objectives. Everything we do in any business or organization is a process.

Strategic planning for excellence

The overall framework for total organizational excellence is shown in Figure 1.1. It all starts with the vision, goals, strategies and mission which should be fully thought through, agreed, and shared in the business. What follows determines whether these are achieved. The factors which are critical to success, the CSFs – the building blocks of the mission – are then be identified. The key performance indicators (KPIs), the measures associated with the CSFs, tell us whether we are moving towards or away from the mission, or just standing still.

Having identified the CSFs and KPIs, the organization should know what are its *core processes*. This is an area of potential bottleneck for many organizations because, if the core processes are not understood, the rest of the framework is difficult to implement. If the processes are known, we can carry out process analysis, sometimes called mapping and flow charting, to fully understand our business and identify opportunities for improvement. By the way, ISO9000 standard based systems should drop out at this stage, rather than needing a separate and huge effort and expense.

Self-assessment to the European (EFQM) Excellence Model or Baldrige Quality model, and benchmarking, will identify further improvement oppor-

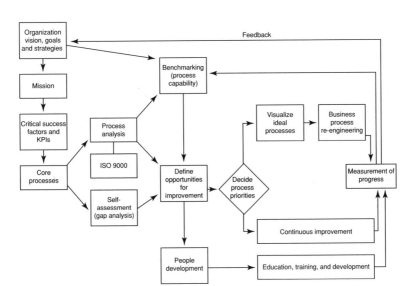

Figure 1.1 The framework for total organizational excellence

tunities. This will create a very long list of things to attend to, many of which require people development, training and education. What is clearly needed next is prioritization – to identify those processes which are run pretty well – they may be advertising/promoting the business or recruitment/selection processes, and subject them to a continuous improvement regime. For those processes which we identify as being poorly carried out, perhaps forecasting, training, or even financial management, we may subject them to a complete re-visioning and re-design activity. That is where BPR comes in. What must happen to all processes, of course, is performance measurement, the results of which feed back to our benchmarking and strategic planning activities.

World-class organizations, of which there need to be more in most countries, are doing **all** of these things. They have implemented their version of the framework and are achieving world-class performance and results. What this requires first, of course, is world-class leadership and commitment.

Chapter 2

Leadership, commitment and culture _____

Key points

This chapter explains how total organizational excellence involves comprehensively improving competitiveness, effectiveness and flexibility through planning, organizing and understanding each activity, and involving each individual at each level. It is achievable in all types of organizations.

Achieving organizational excellence often requires a mind-set change to break down existing barriers, but it must start at the top where the serious obsessional commitment and leadership must be demonstrated. The chief executive must accept the responsibility for commitment to organizational excellence which focuses on the customer needs, but middle management also have a key role to play in communicating the changes required.

The culture of an organization is formed by the beliefs, behaviours, norms, dominant values, rules and the climate in the organization. Total organizational excellence will influence the culture and provide a vision framework comprising its guiding philosophy, core values and beliefs, a purpose, and a mission.

The effectiveness of an organization depends on the extent to which people perform their roles and move towards the common goals and objectives. This chapter is concerned with moving the focus of control from outside individuals to within, so that everyone is accountable for their own performance.

Effective leadership starts with the chief executive's vision and develops into a strategy for implementation. Top management must then develop the following to be effective leaders: clear beliefs and objectives in the form of a mission statement; clear and effective strategies and supporting plans; the critical success factors and core processes; the appropriate management structure; employee participation through empowerment and the continuous improvement cycles.

The guiding principles for senior management are explained in this chapter. Effective leadership is the key to organizational excellence, and this may be demonstrated through:

- commitment to constant improvement,
- a right-first-time philosophy,
- training people to understand customer-supplier relationships,
- not buying on price alone,
- managing systems improvement,
- modern supervision and training,
- managing processes through teamwork and improved communications,
- elimination of barriers and fear,
- constant education and 'expert' development, and
- a systematic approach for integration into the business strategy.

Commitment and policy

To be successful in promoting business efficiency and effectiveness, any approach should be truly organization-wide and it must start at the top with the chief executive, or equivalent. The most senior directors and management should all demonstrate that they are serious about achieving world-class performance. The middle management have a particularly important role to play too, since they should not only grasp the principles, they must go on to explain them to the people for whom they are responsible, and ensure that their own commitment is communicated. Only then will the approach spread effectively throughout the organization. This level of management should also ensure that the efforts and achievements of their subordinates obtain the recognition, attention and reward that they deserve.

The chief executive of an organization must accept the responsibility for and commitment to a policy and strategy in which he/she really believes. This in turn generates responsibilities for a chain of interactions between all the functions: the marketing, design, production/operations, purchasing, distribution and service. Within each and every department of the organization at all levels, starting at the top, basic changes of attitude will be required for organizational excellence. If the owners or directors of the organization do not recognize and accept their responsibilities for the initiation and operation of a total quality approach, then these changes will not happen. Controls, systems and techniques are very important, but they are not the primary requirement for organizational excellence. It is more an attitude of mind, based on pride in the job and teamwork, and it requires total commitment from the management, which then needs to be extended to all employees at all levels and in all departments.

Senior management commitment must be obsessional, not lip service. It is possible to detect real commitment, it shows on the shop floor, in the offices, in the hospital wards – at the point of operation. Going into organizations sporting poster campaigning for excellence instead of belief, one is quickly able to detect the falseness. The people are told not to worry if problems arise, 'just do the best you can', 'the customer will never notice'. The contrast of an organization where total quality means something can be seen, heard, felt. Things happen at this operating interface as a result of **real** commitment. Material problems are corrected with suppliers, equipment difficulties are put right by improved maintenance programmes or replacement, people are trained, change takes place, partnerships are built, continuous improvement is achieved.

The policy

A sound business policy, together with the organization and facilities to put it into effect, is a fundamental requirement, if an organization is to begin the journey to excellence. For example, every organization should develop and state its policy on quality, together with arrangements for its implementation, and the contents of the policy should be made known to all employees. The preparation and implementation of a properly thought out business policy, together with continuous monitoring, makes for smoother production or service operation, minimizes errors and reduces waste.

Management must be dedicated to ongoing improvement, not simply a one-step improvement to an acceptable plateau. These ideas should be set out in the business policy which requires top management to:

a) establish an 'organization'

b) identify the customer

c) identify the customer's needs and perception of needs

d) assess the ability of the organization to meet these needs economically

e) ensure that acquired materials and services reliably meet the required standards of performance and efficiency

f) concentrate on a philosophy of prevention rather than detection of problems

g) educate and train for improvement

h) review the management systems to maintain progress

The business policy must be publicized and understood at all levels of the organization.

Creating or changing the culture

The culture within an organization is formed by a number of components:

- behaviours based on people interactions,
- norms resulting from working groups,
- dominant values adopted by the organization,
- rules of the game for 'getting on',
- the climate.

Culture may be defined then as the beliefs which pervade the organization about how business should be conducted, and how employees should behave and should be treated. Any organization needs a vision framework which includes its **guiding philosophy**, containing the **core values and beliefs**, and a **purpose**. These should be combined into a **mission** which provides a vivid description of what the organization would like to achieve. The **strategies and plans** suggest how it is going to be achieved (Figure 2.1).

The guiding philosophy drives the organization and is shaped by the leaders through their thoughts and actions. It should reflect a visionary organization

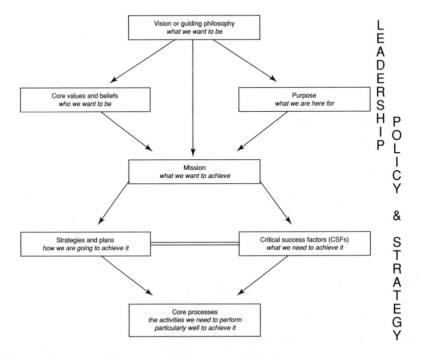

Figure 2.1 Vision framework for an organization

rather than the vision of a single leader, and should evolve with time, although organizations must hold on to the **core** elements.

The core **values** and beliefs represent the organization's basic principles about what is important in business, its conduct, its social responsibility and its response to changes in the environment. They should act as a guiding force, with clear and authentic values, which are focused on employees, suppliers, customers, society at large, safety, shareholders, and generally stakeholders. The Exxon Chemical version of this is set into indestructible polyethylene (Figure 2.2)!

The **purpose** of the organization should be a development from the core values and beliefs and should quickly and clearly convey what is the organization's role.

The **mission** will translate the abstractness of philosophy into tangible goals that will move the organization forward and make it perform to its optimum. It should not be limited by the constraints of strategic analysis, and should be proactive not reactive. Strategy is subservient of mission – the strategic analysis being done after, not during, the mission setting process.

Control

The effectiveness of an organization and its people depends on the extent to which each perform their role and move towards the common goals and objectives. Control is the process by which information or feedback is provided so as to keep all functions on track, being the sum total of the activities which increase the probability that the planned results will be achieved. Control mechanisms fall into three categories, depending upon their position in the managerial process (Table 2.1).

Table 2.1

Before the fact	Operational	After the fact
Strategic plan	Observation	Annual reports
Action plans	Inspection and correction	Variance reports
Budgets	Progress review	Audits
Job descriptions	Staff meetings	Surveys
Individual performance objectives	Internal information and data systems	Performance review
Training and development	Training programmes	Evaluation of training

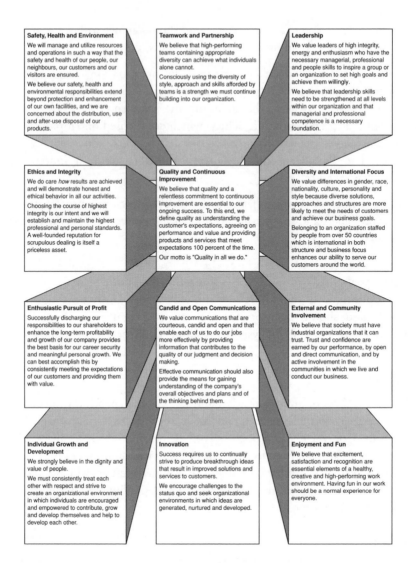

Figure 2.2 Exxon Chemical's leadership commitment

Many organizations use only after-the-fact controls, causing managers to take a reactive rather than a proactive position. Such 'crisis-orientation' needs to be replaced by a more anticipative one in which the focus is on preventive or before the fact controls.

Attempting to control performance through systems, procedures or techniques **external** to the individual is not an effective approach since it relies on 'controlling' others; individuals should be responsible for their own actions.

An externally-based control system can result in a high degree of concentrated effort in a specific area if the system is overly structured, but it can also cause negative consequences to surface:

i)   Since all rewards are based on external measures, which are imposed, the 'team members' often focus all their efforts on the measure itself, i.e., to have it set lower (or higher) than possible, to manipulate the information which serves to monitor it, or to dismiss it as someone else's goal not theirs. In the budgeting process, for example, distorted figures are often submitted by those who have learned that their 'honest projections' will be automatically altered anyway.

ii) When the rewards are dependent on only one or two limited targets, all efforts are directed at those, even at the expense of others. If short-term profitability is the sole criterion for bonus distribution or promotion, it is likely that investment for longer term growth areas will be substantially reduced by those involved. Similarly, strong emphasis and reward for output or production may result in lowered quality.

iii) The fear of not being rewarded, or even being criticized, for performance that is less than desirable, may cause some to withhold information that is unfavourable but nevertheless should be flowing into the system.

iv) When reward and punishment is used to motivate performance, the degree of risk taking may lessen and be replaced by a more cautious and conservative approach. In essence, the fear of failure replaces the desire to achieve.

The following situations have been observed by the author and his colleagues within companies which have taken part in research and consultancy during the last few years:

1   Goals are imposed which are seen or known to be unrealistic. If the goals perceived by the subordinate are in fact accomplished then the subordinate has proved himself wrong. This clearly has a negative effect on the effort expended, since few people are motivated to prove themselves wrong!

2   Where individuals are stimulated to commit themselves to a goal and, where their personal pride and self-esteem are at stake, then the level of motivation is at a peak. For most people the toughest critic and the hardest taskmaster they confront is not their immediate boss, but themselves.

3   Directors and managers are often afraid of allowing subordinates to set the goals for fear of them being set too low, or loss of control over subor-

dinate behaviour. It is also true that many do not wish to set their own targets but prefer to be told what is to be accomplished.

The approach found in most world-class organizations is concerned with moving the focus of control from outside the individual to within; the objective being to make everyone accountable for their own performance, and to get them committed to achieving in a highly motivated fashion. The assumptions a director or manager must make in order to move in this direction are simply that people do not need to be coerced to perform well, and that people want to achieve, accomplish, influence activity, and challenge their abilities. If there is belief in this, then only the techniques remain to be discussed.

Total organizational excellence is user driven, it cannot be imposed from outside the organization, as can perhaps a quality standard or statistical process control. This means that the ideas for improvement must come from those with knowledge and experience of the processes, activities and tasks, and this has massive implications for training and follow-up. Organizational excellence is not a cost-cutting or productivity improvement device in the traditional sense, and it must not be used as such, although the effects of its successful implementation will certainly reduce costs and improve productivity. It is concerned chiefly with changing attitudes and skills so that the culture of the organization becomes one of preventing failure – doing the right things, right first time, every time.

Effective leadership

Some management teams have broken away from the traditional style of management, they have made a 'managerial breakthrough'. Their new approach puts their organizations head and shoulders above competitors in the fight for sales, profits, resources, funding and jobs. Many public service organizations are beginning to move in the same way and the successful strategy they are adopting depends very much on effective leadership.

Effective leadership starts with the chief executive's vision, capitalizing on market or service opportunities, continues through a strategy which will give the organization competitive advantage, and leads to business or service success. It goes on to embrace all of the beliefs and values held, the decisions taken and the plans made by anyone anywhere in the organization and the focusing of them into effective, value-adding, action.

Key leadership points

There are five main things which top management must do to be effective leaders:

1   **Develop and publish clear documented corporate beliefs and purpose - *a mission statement*.** Executives must express values and beliefs through a clear vision of what they want their company or organization to be and its purpose - what they specifically want to achieve in line with the basic beliefs. Together, they define what the company or organization is all about. The senior management team will need to spend some time away from the 'coal face' to do this and develop their programme for implementation.

Clearly defined and properly communicated beliefs and purpose, which can be summarised in the form of a mission statement, are essential if the directors, managers and other employees are to work together as a winning team.

The beliefs and purpose should address:

- The definition of the business – for example, the needs that are satisfied or the benefits provided.
- A commitment to effective leadership and quality.
- Target sectors and relationships with customers, and market or service position.
- The role or contribution of the company, organization, or unit – for example, profit generator, service department, opportunity seeker.
- The distinctive competence – a brief statement which applies only to that organization, company or unit.
- Indications for future direction – a brief statement of the principal plans which would be considered.
- Commitment to monitoring performance against customers' needs and expectations, and continuous improvement.

The mission statement and the broader beliefs and purpose may then be used to communicate an inspiring vision for the organization of where it is going. The top management must then show total commitment to it.

2   **Develop clear and effective strategies and supporting plans for achieving the mission.** The achievement of the company or service mission requires the development of business or service strategies, including the strategic positioning in the 'market place'. Plans can then be developed for implementing the strategies. Such strategies and plans can be developed by senior managers alone, but there is likely to be more commitment to them if employee participation in their development and implementation is encouraged.

3 **Identify the critical success factors and core processes** (Figure 2.1). The next step is the identification of the **critical success factors** (CSFs), a term used to mean the most important sub-goals of a business or organization. CSFs are what must be accomplished for the mission to be achieved. The CSFs are followed by the **core business processes** for the organization – the activities which must be done particularly well for the CSFs to be achieved.

This process is described in more detail in Chapter 3, Strategic planning.

4 **Review the management structure.** Defining the corporate mission, strategies, CSFs and core processes might make it necessary to review the organizational structure. Directors, managers and other employees can be fully effective only if an effective structure based on process management exists. This includes both the definition of responsibilities for the organization's management and the operational procedures they will use. These must be the agreed best ways of carrying out the core processes.

The review of the management structure should also include the establishment of a process team structure throughout the organization.

5 **Empowerment – encourage effective employee participation.** For effective leadership, it is necessary for management to get very close to the employees. They must develop effective communications – up, down and across the organization – and take action on what is communicated; encourage good communications between all suppliers and customers.

Particular attention must be paid to:

Attitudes – The key attitude for managing any winning company or organization may be expressed as: 'I will personally understand who my customers are and what are their needs and expectations of me; I will measure how well I am satisfying their needs and expectations and I will take whatever action is necessary to satisfy them fully. I will also understand and communicate my requirements to my suppliers, inform them of changes and provide feedback on their performance'.

This attitude must start at the top – with the chairman or chief executive. It must then percolate down to be adopted by each and every employee. That will only happen if managers lead by example. Words are cheap and will be meaningless if employees see from managers' actions that they do not actually believe or intend what they say.

Abilities – Every employee must be able to do what is needed and expected of him, or her – but it is first necessary to decide what is really needed and expected. If it is not clear what the employees are required to do and what the standards of performance are expected, how can managers expect them to do it?

Educate, train, educate, train, educate and train again. Education and training is very important, but can be expensive if the money is not spent wisely. The education, training and development must be related to needs, expectations, and process improvement. It must be planned and **always** its effectiveness reviewed.

Participation – If all employees are to participate in making the company or organization successful (directors and managers included), then they must also be trained in the basics of disciplined management. They must be trained to:

E   **Evaluate** – the situation and define their **objectives,**

P   **Plan** – to fully achieve those objectives,

D   **Do** – implement the plans,

C   **Check** – that the objectives are being achieved, and

A   **Amend** – take corrective action if they are not.

The word 'disciplined' means that people, at all levels, must do what they say they will do. It also means that, in whatever they do, they will go through the full process of evaluate, plan, do, check and amend, rather than the more traditional and easier option of starting by doing rather than evaluating. This will lead to a never-ending improvement helix (Figure 2.3).

This basic approach needs to be backed up with good project management, planning techniques and problem solving methods, which can be taught to anyone in a relatively short period of time. The project management enables changes to be made successfully and the problem solving helps people to remove the obstacles in their way. Directors and managers need this training as much as other employees.

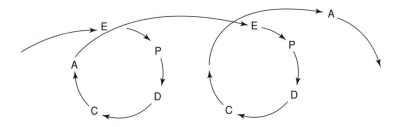

Figure 2.3   The helix of never-ending improvement

Ten principles for senior management

The vehicle for achieving effective leadership is understanding and commitment. We have seen that it must involve the entire organization, all the people and all the functions, including external organizations and suppliers. In the first two chapters, several facets of organizational excellence have been identified:

- Management's responsibility for setting the guiding philosophy, vision, policy, etc., and providing motivation through leadership and equipping people to achieve.

- Recognizing customers and discovering their needs.

- Controlling processes, including systems, and improving their capability.

- Empowerment of people at all levels in the organization to make improvements.

The idea of total organizational excellence can be daunting and the chief executive and directors contemplating it may become confused and irritated by the proliferation of theories and packages. A simplification is required. The **core** of excellence must be the customer–supplier interfaces, both internally and externally, and the fact that, at each interface there are processes which convert inputs to outputs. Clearly, there must be commitment to building-in quality through management of the inputs and processes.

How can senior managers and directors be helped in their understanding of what needs to be done to become committed to achieve organization excellence and implement the vision? Presented below are ten principles for senior management to consider. They may not lead directly to excellence, but they certainly can be found in those organizations which aspire to become world-class.

1 **The organization needs long term *commitment* to constant improvement.** There must be a constancy of purpose, and commitment to it must start from the top. The improvement process must be planned on a truly organization-wide basis, i.e., it must embrace all locations and departments and must include customer, suppliers, and sub-contractors. It cannot start in 'one department' in the hope that the programme will spread from there.

The place to start is in the boardroom – leadership must be by example. Then **progressively** expand it to embrace all parts of the organization. It is wise to avoid the 'blitz' approach which can lead to a lot of hype but no real changes in behaviour.

2 **Adopt the philosophy of zero errors/defects to change the *culture* to right first time.** This must be based on a thorough understanding of the customer's needs and expectations and teamwork, developed through employee participation and rigorous application of the EPDCA helix.

3 **Educate people to understand the *customer–supplier* relationships.** Again the commitment to the customer needs must start from the top, from the chairman or chief executive. Without that time and effort will be wasted. Customer orientation must then be achieved for each and every employee, directors and managers. The concept of internal customers and suppliers must be thoroughly understood and used.

4 **Do not buy products or services on price alone - look at the *total cost*.** Demand continuous improvement in everything, including suppliers. This will bring about improvements in product, service and failure rates. Continuously improve the product or the service provided externally, so that the total costs of doing business are reduced.

5 **Recognize the improvement of the *systems* needs to be managed.** Defining the performance standards expected and the systems to achieve them is a managerial responsibility. The rule has to be that the systems will be in line with the shared needs and expectations and will be part of the continuous improvement process.

6 **Adopt modern methods of *supervision* and *training* – eliminate fear.** It is all too easy to criticise mistakes, but it often seems difficult to praise efforts and achievements. Recognize and publicize efforts and achievements and provide the right sort of facilitation, supervision, and training.

7 **Eliminate barriers between departments by managing the *processes* – improve *communications* and *teamwork*.** Barriers are often created by 'silo management', in which departments are treated like containers which are separate from one another. The customers are not interested in departments – they stand on the outside of the organization and see slices through it – they see the **processes**. It is necessary to build teams and improve communications around the processes.

8 **Eliminate:**

   • **arbitrary goals without methods,**

   • **all standards based only on numbers,**

   • **barriers to pride of workmanship, and**

   • **fiction, get *facts* by using the correct *tools*.**

At all times it is essential to know how well you are doing in terms of satisfying the customers' needs and expectations. Help every single employee to know **how** they will achieve the goals and how well they are doing.

Traditional 'piecework' will not survive in a successful organization because it creates barriers and conflict. People should be proud of what they do and not be encouraged to behave like monkeys being thrown peanuts.

Train people to measure and report performance in language that the people doing the job can understand. Encourage each employee to measure his/her own performance. Do not stop with measuring performance in the organization – find out how well other organizations (competitive or otherwise) are performing against similar needs and expectations – **benchmark** against best practice.

The costs of mismanagement and the level of fire-fighting together are excellent factual indicators of the internal health of an organization. They are relatively easily measured and simple for most people to understand.

9   **Constantly educate and retrain – develop the '*experts*' in the business.** The experts in any business are the people who do the job every day of their lives. The 'energy' which lies within them can be released into the organization through education, training, encouragement and involvement.

10 **Develop a *systematic* approach to organizational excellence.** Total organizational excellence should not be regarded as a woolly-minded approach to running a business. It requires a carefully planned and fully integrated strategy, derived from the mission. That way it will help any organization to realise its vision.

In summary

- identify **customer/supplier** relationships

- manage **processes**

- change the **culture**

- improve **communication**

- show **commitment**

This provides a multi-dimensional 'vision' against which a particular company's status can be examined, or against which a particular approach may be compared and weaknesses highlighted. One of the greatest tangible benefits of this sort of leadership is the increased market share or 'results', rather than just the reduction in costs. The evidence for this can already be seen in some of the major consumer and industrial markets of the world. Superior performance can be converted into premium prices – quality clearly correlates with profit. The less tangible benefit of greater employee involvement is equally, if not more, important in the longer term. The pursuit of continual improvement must become a way of life for everyone in any organization if it is to succeed in today's competitive environment.

(The reader will find further material on leadership and people development in Chapter 10.)

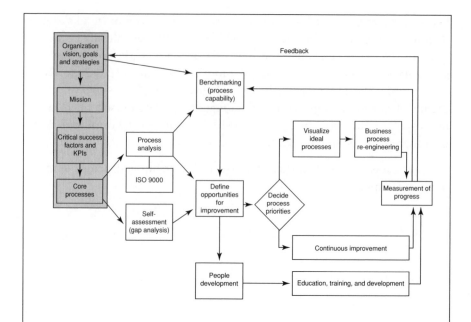

Strategic planning _____

Chapter 3

Strategic planning _____

Key points

Senior managers in many different organizations recognize the need for better planning to deal with increasing competitiveness, calls for more efficiency, pressure on costs, etc., but lack an understanding of how to implement the changes. Total organizational excellence is effected not by focusing on formal structures and systems, but by aligning process management teams. The chapter explains how strategic planning starts with writing the mission statement, analysis of the critical success factors (CSFs) and understanding the core processes.

Some of the obstacles to organizational excellence and resistance to change may be overcome through education, communication, participation, involvement, facilitation, and support. The 'blitz' or rapid change approach should be rejected in favour of a slow, planned purposeful one, starting at the top.

This chapter should help senior management begin the task of process alignment through seven steps to a self-reinforcing cycle of commitment, communication, and culture change:

1 gain commitment to change;
2 develop a shared vision and mission of the business or desired change;
3 develop the mission into its CSFs;
4 define the key performance indicators and targets;
5 understand the core processes; break down the core processes into sub-processes, activities and tasks, monitoring and adjusting the process-alignment in response to changes; and
6 goal translation which then ensures that the 'whats' are converted into 'hows' using a quality function deployment matrix based process to achieve total organization excellence.

The management of change

I recall the managing director of a large petrochemical company who decided that a major change was required in the way the company operated if serious competitive challenges were to be met. A human resources director was recruited who was given the task of managing the change in the people and their 'attitudes'. After several 'programmes' aimed at achieving the required change, including a new structure for the organization, a staff appraisal system linked to pay, and seminars to change attitudes, very little change in actual organizational behaviour had occurred.

Clearly something had gone wrong somewhere, but what, who, where? Everything was wrong, including what needed changing, who should lead the changes, and in particular, how the changes should be brought about. This type of problem is very common in organizations which desire to change the way they operate to deal with increased competition, a changing market place, and different business rules.

In this situation many executives recognize the need to move away from an autocratic management style, with formal rules and hierarchical procedures, and narrow work demarcations – some have tried to create teams, to delegate (perhaps for the first time), and to improve communications. But they lack an understanding of how the change should be implemented. They often believe that changing the formal organizational structure, having 'culture change' programmes, training courses, and new payment systems will, by themselves, make the transformations.

In much research work carried out at the European Centre for Business Excellence, the research division of Oakland Consulting, it has been shown that there is almost an inverse relationship between successful change and having formal organization-wide 'change programmes'. This is particularly true if one functional group, such as HR, 'owns' the programme.

In several large organizations in which total quality management has been used successfully to effect change, the senior management did not focus on formal structures and systems, but set up process management teams to solve real business or organization problems. The key to success in this area is to align the employees of the business, their roles and responsibilities with the organization and its processes. This is the core of process mapping or total organizational alignment. When an organization focuses on its core processes, that is the activities and tasks themselves, rather than on abstract issues such as 'culture' and 'participation', then the change process can begin in earnest.

An approach to change based on process alignment, starting with the mission statement, analysing the critical success factors, and moving on to the core processes is the most effective way to engage the organization's people in an enduring change process. Many change programmes do not work because they begin trying to change the knowledge, attitudes and beliefs of individuals. The theory is that changes in these areas will lead to changes in behaviour throughout the organization. It relies on a form of religion spreading through the people in the business.

What is often required, however, is virtually the opposite process, based on the recognition that people's behaviour is determined largely by the roles they have to take up. If we create for them new responsibilities, team roles, and a process driven environment, a new situation will develop that will force their attention and work on the processes. This will change the culture. Teamwork is an especially important part of the model, in terms of bringing about change. If changes are to be made in quality, costs, market, product or service development, close co-ordination among the marketing, design, production/operations and distribution groups is essential. This can be brought about effectively only by multi-functional teams working on the processes and understanding their interrelationships. **Commitment** is a key element of support for the high levels of co-operation, initiative, and effort which will be required to understand and work on the labyrinth of processes which exist in most organizations. In addition to the knowledge of the business as a whole, which will be brought about by an understanding of the mission, CSFs, process breakdown links, certain tools, techniques, and interpersonal skills will be required for good **communication** around the processes. These are essential for people to identify and solve problems as teams.

If any of these elements are missing, the change process will collapse. The difficulties experienced by many organizations' formal change processes is that they tackle only one or two of these necessities. Many organizations which try to create a new philosophy based on teamwork, fail to recognize that the employees do not know which teams to form or how they should function as teams. Recognition that effective teams need to be formed around a **process,** which they begin to understand together – perhaps for the first time – and then be helped as individuals through the forming–storming–norming–performing sequence will generate the interpersonal skills and attitude changes necessary to make the new 'structure' work (see Chapters 5 and 10 for further details).

Obstacles to implementation

Some of the obstacles to achieving total organizational excellence are that it can be seen as time-consuming, bureaucratic, formalistic, rigid, impersonal, and/or the property of a specialist group. There is frequently found the so-called middle management resistance, particularly in organizations where there is a fear of openness.

Some of this is typical resistance to any change. This may be more severe if the organization is successful, if there is a particularly deep-seated culture, if there has been a great deal of change already, or if the change lacks legitimacy.

The methods of overcoming resistance to change are largely the subject of this book, but they include:

- education and communication

- participation and involvement

- facilitation and support

- negotiation and agreement

For successful change to occur, of course, there must be the perceived need for it, appropriate resources, and a supportive organizational culture.

Integrating organizational excellence into the strategy of the business

Organizations will avoid the problems of 'change programmes' by concentrating on 'process alignment' – recognizing that people's roles and responsibilities must be related to the processes in which they work. Senior managers may begin the task of process alignment by a series of seven steps which are distinct but clearly overlap. This recommended path develops a self-reinforcing cycle of **commitment, communication**, and **culture** change. The order of the steps is important because some of the activities will be inappropriate if started too early. In the initial stages of managing change, timing can be critical.

Step one   *Gain commitment to change through the organization of the top team*

Process alignment requires the starting point to be a broad review of the organization and the changes required by the top management team. By gaining this shared diagnosis of what an organization is required to change, what the 'business' problems are and/or what must be improved, the most senior executive mobilizes the initial commitment that is vital to begin the change process. An important element here is to get the top team working as a team and techniques such as MBTI and/or FIRO-B will play an important role (see Chapter 10).

At this stage of the process, strong leadership from the top is crucial. Commitment to the change, whatever it may be, is always imbalanced. Some senior managers may be antagonistic, some neutral, others enthusiastic or worried about the proposed changes.

Step two   *Develop a shared vision and mission for the business, or of what change is required*

Once the top team is committed to the analysis of the changes required, it can develop vision and mission statements which will help to define the new process-alignment, roles and responsibilities. This will lead to a co-ordinated flow of analysis of process that crosses the traditional functional areas at all levels of the organization, without changing formal structures, titles, and systems which can create resistance.

The mission statement gives a purpose to the organization or unit. It should answer the questions 'what are we here for?' or 'what is our basic purpose?' and 'what have we got to achieve?' It therefore defines the boundaries of the business in which the organization operates. This will help to focus on the 'distinctive competence' of the organization, and to orient everyone in the same direction of what has to be done. The mission must be documented, agreed by the top management team, sufficiently explicit to enable its eventual accomplishment to be verified, and ideally be no more than four sentences. The statement must be understandable, communicable, believable, and usable.

The mission statement is:

- an expression of the aspiration of the organization,
- the touchstone against which all actions or proposed actions can be judged,
- usually long term,
- short term if the mission is survival.

Typical content includes a statement of:

- The role or contribution of the business or unit – for example, profit generator, service department, opportunity seeker.
- The definition of the business – for example, the needs you satisfy or the benefits you provide. Do not be too specific or too general.
- Your distinctive competence – this should be a brief statement that applies to only your specific unit. A statement which could apply equally to any organization is unsatisfactory.
- Indications for future direction – a brief statement of the principal things you would give serious consideration to.

Some questions that may be asked of a mission statement are, does it:

- define the organization's role?
- contain the need to be fulfilled:
  - is it worthwhile/admirable?
  - will employees identify with it?

    – how will it be viewed externally?

- take a long-term view, leading to, for example, commitment to new product or service development, or training of personnel?

- take into account all the 'stakeholders' of the organization?

- ensure the purpose remains constant despite changes in top management?

It is important to establish in some organizations whether or not the mission is survival. This does not preclude a longer-term mission, but the short-term survival mission must be expressed, if it is relevant. The management team can then decide whether they wish to continue long-term strategic thinking. If survival is a real issue the author and his colleagues would advise against concentrating on the long-term planning initially.

There must be open and spontaneous discussion during generation of the mission, but there must in the end be convergence on one statement. If the mission statement is wrong, everything that follows will be wrong too, so a clear understanding is vital.

Step three   *Develop the 'mission' into its critical success factors (CSFs) to coerce and move it forward*

The development of the mission is clearly not enough to ensure its implementation. This is the 'danger gap' which many organizations fall into because they do not foster the skills needed to translate the mission through its CSFs into the core processes. Hence, they have 'goals without methods' and change is not integrated properly into the business.

Once the top managers begin to list the CSFs they will gain some understanding of what the mission or the change requires. The first step in going from mission to CSFs is to brainstorm all the possible impacts on the mission. In this way 30 to 50 items ranging from politics to costs, from national cultures to regional market peculiarities may be derived.

The CSFs may now be defined – *what* the organization must accomplish to achieve the mission, by examination and categorization of the impacts. This should lead to a balanced set of deliverables for the organization in terms of:

- financial and non-financial performance;

- customer/market satisfaction;

- people/internal organization satisfaction;
- environmental/societal satisfaction.

There should be no more than eight CSFs, and no more than four if the mission is survival. They are the building blocks of the mission – minimum key factors or sub-goals that the organization **must have** or **needs** and which together will achieve the mission. They are the *whats* not the *hows*, and are not directly manageable – they may be in some cases statements of hope or fear. But they provide direction and the success criteria, and are the end product of applying the processes. In CSF determination, a management team should follow the rule that each CSF is **necessary** and together they are **sufficient** for the mission to be achieved.

Some examples of CSFs may clarify their understanding:

- We must have right-first-time suppliers.
- We must have motivated, skilled workers.
- We need new products that satisfy market needs.
- We need new business opportunities.
- We must have best-in-the-field product quality.

The list of CSFs should be an agreed balance of strategic and tactical issues, each of which deals with a 'pure' factor, the use of 'and' being forbidden. It will be important to know when the CSFs have been achieved, but an equally important step is to use the CSFs to enable the identification of the processes.

Senior managers in large complex organizations may find it necessary or useful to show the interaction of divisional CSFs with the corporate CSFs in an impact matrix (see Figure 3.1 and discussion under Step seven).

Step four   *Define the key performance indicators as being the quantifiable indicators of success in terms of the mission and CSFs*

The mission and CSFs provide the **what** of the organization, but they must be supported by measurable key performance indicators (KPIs) that are tightly and inarguably linked. These will help to translate the directional and sometimes 'loose' statements of the mission into clear **targets**, and in turn to simplify management's thinking. The KPIs will be used to monitor progress and as evidence of success for the organ-

Figure 3.1 Interaction of corporate and divisional CSFs

ization, in every direction, internally and externally.

Each CSF should have an 'owner' who is a member of the management team that agreed the mission and CSFs. The task of an owner is to:

- define and agree the KPIs and associated *targets;*

- ensure that appropriate data is collected and recorded;

- monitor and report progress towards achieving the CSF (KPIs and targets) on a regular basis;

- review and modify the KPIs and targets where appropriate

A typical CSF data sheet for completion by owners is shown in Figure 3.2.

Step five   *Understand the core processes and gain process sponsorship*

This is the point when the top management team have to consider how to institutionalize the mission or the change in the form of processes that will continue to be in place, after any changes have been effected.

The core business processes describe what actually is or needs to be done so that the organization meets its CSFs. As with the CSFs and the mission, each process which is **necessary** for a given CSF must be identified, and together the processes listed must be **sufficient** for all the CSFs to be accomplished. To ensure that **processes** are listed, they should be in the form of verb plus object, such as research the market, recruit competent staff, or manage supplier performance. The core processes identified frequently run across 'departments' or functions, yet they must be measurable.

Each core process should have a sponsor, preferably a member of the management team that agreed the CSFs.

The task of a sponsor is to:

- ensure that appropriate resources are made available to map, investigate and improve the process;

- assist in selecting the process improvement team leader and members;

- remove blocks to the teams' progress;

- report progress to the senior management team.

CSF No.	We must have / we need

CSF Owner

Key performance indicators (KPIs)

Core processes impacting on this CSF

Process No.	Process	Impacts on other CSFs	Process performance	Agreed sponsor

Figure 3.2 CSF data sheet

The first stage in understanding the core processes is to produce a set of processes of a common order of magnitude. Some smaller processes identified may combine into core processes, others may be already at the appropriate level. This will ensure that the change becomes entrenched, the core processes are identified and that the right people are in place to sponsor or take responsibility for them. This will be the start of getting the process team organization up and running.

The questions will now come thick and fast; is the process currently carried out? By whom? When? How frequently? With what performance and how well compared with competitors? The answering of these will force process ownership into the business. The process sponsor may form a process team which takes quality improvement into the next steps. Some form of prioritization using process performance measures is necessary at this stage to enable effort to be focused on the key areas for improvement. This may be carried out by a form of impact matrix analysis (see Figure 3.3). The outcome should be a set of 'most critical processes' (MCPs) which receive priority attention for improvement, based on the number of CSFs impacted by each process and its performance on a scale A to E.

Step six Break down the core processes into sub-processes, activities and tasks and form improvement teams around these

Once an organization has defined and mapped out the core processes, people need to develop the skills to understand how the new process structure will be analysed and made to work. The very existence of new process teams with new goals and responsibilities will force the organization into a learning phase. The changes should foster new attitudes and behaviours.

An illustration of the breakdown from mission through CSFs and core processes, to individual tasks may assist in understanding the process required:

Mission

Two of the statements in a well known management consultancy's mission statement are:

'Gain and maintain a position as Europe's foremost management consultancy in the development of organizations through management of change.

Provide the consultancy, training and facilitation necessary to assist with making continuous improvement an integral part of our customers' business strategy.'

No.	Process	CSF No.								Number of CSF impacts	A-E ranking

A-E process ranking: A-Excellent; B-Good; C-Average; D-Poor; E-Embryonic

Figure 3.3 Process/CSF matrix

Ctitical Success Factor

One of the CSFs which clearly relates to this is:

We need a high level of awareness of our company in the market place

Core Process

One of the core processes which clearly must be done particularly well to achieve this CSF is to:

Promote, advertise, and communicate the company's business capability.

Sub Process

One of the sub-processes which results from a breakdown of this core process is:

Prepare the company's information pack.

Activity

One of the activities which contributes to this sub-process is:

Prepare **one** of the subject booklets, i.e. 'Business Excellence and Self-Assessment'.
↓

Task

One of the tasks which contributes to this is:

Write the detailed leaflet for a particular seminar, e.g. one or three day seminars on self-assessment.

Individuals, tasks and teams

Having broken down the processes into sub-processes, activities and tasks in this way, it is now possible to link this with the Adair model of action centred leadership and teamwork (see Chapter 10).

The **tasks** are clearly performed, at least initially, by individuals. For example, some**body** has to sit down and draft out the first version of a seminar leaflet. There has to be an understanding by the individual of the task and its position in the hierarchy of processes. Once the initial task has been performed, the results must be referenced to the overall activity of co-ordinating the promotional booklet – say for business excellence. This clearly brings in the team role and there must be interfaces between the needs of the **tasks**, the **individuals** who performed them and the **team** concerned with the **activities**.

Using the hierarchy of processes, it is possible to link this with the hierarchy of teams. Hence:

Senior management team:	Mission–CSFs–core processes
Process teams:	Core processes
Improvement (or functional) teams	{Sub-process {Activities
'Local teams'/individuals:	Tasks

Performance measurement and metrics

Once the processes have been analysed in this way, it should be possible to develop **metrics** for measuring the performance of the processes, sub-processes, activities, and tasks. These must be meaningful in terms of the **inputs** and **outputs** of the processes, and in terms of the **customers** of and **suppliers** to the processes (Figure 3.4).

At first thought this can seem very difficult for processes such as preparing a sales brochure or writing leaflets advertising seminars, but if we think carefully about the **customers** for the leaflet writing tasks,

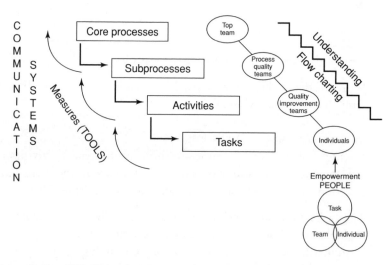

Figure 3.4 Breakdown of core processes into sub-processes, activities and tasks

these will include the **internal** ones, i.e., the consultants – does the output meet their requirements? Does it really say what the seminar is about, what its objectives are and what the programme will be? Clearly, one of the 'measures' of the seminar leaflet writing task could be the number of typing errors on it, but is this a **key** measure of the performance of the process? Only in the context of office management is this an important measure. Elsewhere it is not.

Similarly for the **activity** of preparing the subject booklet. Does it tell the 'customer' what business excellence is and how the consultancy can help? For the **sub-process** of preparing the company brochure, does it inform people about the company and does it bring in enquiries from which can be developed customers? Clearly, some of these measures require **external market research**, some of them require **internal research**. The main point is that metrics must be developed and used which reflect the **true performance** of the processes, sub-processes, activities, and tasks. These must involve good contact with external and internal customers of the processes. The metrics may be quoted as **ratios**, e.g., number of customers derived per number of brochures mailed out. Good data collection, record keeping, and analysis is clearly required.

It is hoped that this illustration will help the reader to:

- Understand the breakdown of processes to sub-processes, activities, and tasks.

- Understand the links between these process breakdowns and the task–individual–team concepts.

- Link the hierarchy of processes with the hierarchy of teams.

- Begin to assemble a cascade of flow charts representing the process breakdowns, which can form the basis of the quality system and communicate what is going on throughout the business.

- Understand the way in which metrics must be developed to measure the true performance of the process, and their links with the customers, suppliers, inputs and outputs of the processes.

The changed patterns of co-ordination, driven by the process maps, should increase involvement, collaboration and information sharing. Clearly the senior and middle managers must provide the right support. Once employees, at all levels, identify what kinds of new skills are needed, they will ask for the formal training programmes in order to develop those skills further. This is a key area because the

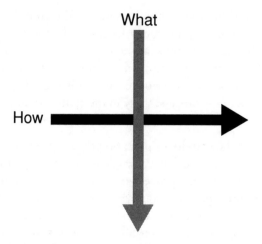

Figure 3.5   The goal translation process

teamwork around the processes will ask more of employees, so they will need increasing support from their managers.

This has been called 'just-in-time' training which describes very well the nature of the training process required. This contrasts with the blanket or carpet bombing training associated with many unsuccessful change programmes, which targets competencies or skills, but does not change the organization's patterns of collaboration and co-ordination.

Step seven   *Ensure process and people alignment through a policy deployment or goal translation process*

One of the keys to integrating excellence into the business strategy is a formal 'goal translation' or 'policy deployment' process. If the mission and measurable goals have been analysed in terms of critical success factors and core processes, then the organization has begun to understand how to achieve the mission. Goal translation ensures that the 'whats' are converted into 'hows', passing this right down through the organization, using a quality function deployment (QFD) type process, Figure 3.5 (see *Total Quality Management*, 2nd edition, by John Oakland). The method is best described by an example.

At the top of an organization in the chemical process industries, five measurable goals have been identified. These are listed under the heading 'What' in Figure 3.6. The top team listen to the 'voice of the customer' and try to understand *how* these business goals will be achieved. They realize that product consistency, on-time delivery, and

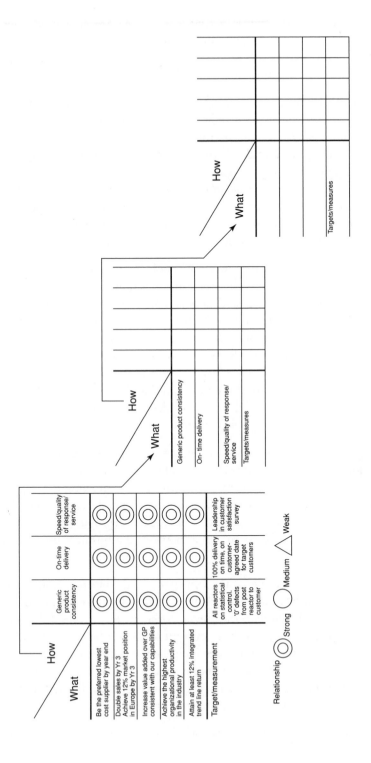

Figure 3.6 The goal translation process

speed or quality of response are the keys. These CSFs are placed along the first row of the matrix and the relationships between the *what* and the *how* estimated as strong, medium or weak. A measurement target for the hows is then specified.

The *how* becomes the *what* for the next layer of management. The top team share their goals with their immediate reports and ask them to determine their *hows*, indicate the relationship, and set measurement targets. This continues down the organization through a 'catch-ball' process until the senior management goals have been translated through the *what/how* → *what/how* → *what/how* matrices to the individual tasks within the organization. This provides a good discipline to support the breakdown and understanding of the business process mapping described previously and in detail in Chapter 5.

A successful approach to policy/goal deployment and strategic planning in an organization with several business units or divisions, is that mission, CSFs with KPIs and targets, and core processes, are determined at the corporate level, typically by the board. Whilst there needs to be some flexibility about exactly how this is translated into the business units, typically it would be expected that the process is repeated with the senior team in each business unit or division. Each business unit head should be part of the top team that did the work at the corporate level, and each of them would develop a version of the same process with which they feel comfortable.

Each business unit would then follow a similar series of steps to develop their own mission (perhaps) and certainly their own CSFs and KPIs with targets. A matrix for each business unit showing the impact of achieving the business unit CSFs on the corporate CSFs would be developed. In other words, the first deployment of the corporate 'whats' CSFs is into the 'hows' – the business unit CSFs (Figure 3.1).

If each business unit follows the same pattern, the business unit teams will each identify unit CSFs, KPIs with targets and core processes, which are interlinked with the ones at corporate level. Indeed the core processes at corporate and business unit level may be the same, with any specific additional processes identified at business unit level to catch the flavour and business needs of the unit. It cannot be over-emphasized how much ownership there needs to be at the business unit management level for this to work properly.

With regard to core processes, each business unit or function will

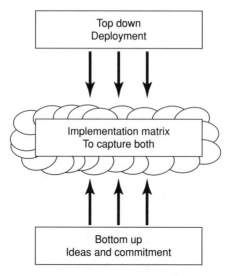

Figure 3.7   Implementation: top down and bottom up approach

begin to map these at the top level. This will lead to an understanding of the purpose, scope, inputs, outputs, controls, and resources for each process and provide an understanding of how the sub-processes are linked together. Flow charting showing connections with procedures will then allow specific areas for improvement to be identified so that the continuous improvement, 'bottom up' activities can be deployed, and benefit derived from the process improvement training to be provided (Figure 3.7).

It is important to get clarity at the corporate and business unit management levels about the whats/hows relationships, but the ethos of the whole process is one of involvement and participation in goal/target setting, based on good understanding of processes – so that it is known and agreed what can be achieved and what needs measuring and targeting at the business unit level.

Senior management may find it useful to monitor performance against the CSFs, KPIs and targets, and to keep track of process using a reporting matrix, perhaps at their monthly meetings. A simplified version of this developed for use in a small company is shown in Figure 3.8. The frequency of reporting for each CSF, KPI, and process can be determined in a business planning calendar.

As previously described, in a larger organization, this approach may be used to deploy the goals from the corporate level through divisions to

Core processes	CSFs: We must have — Satisfactory financial and non-financial performance	A growing base of satisfied customers	A sufficient number of committed and competent people	Research projects properly completed and published	** = Priority for improvement	Process owner	Process performance	Measures and targets
Manage people	X	X	X	X	**			
Develop products		X						
Develop new business	X	X			**			
Manage our accounts	X	X		X	**			
Manage financials	X		X					
Manage int. systems	X	X	X					
Conduct research		X		X				

CSFs: We must have	Measures	Year targets	CSF owner
Satisfactory financial and non-financial performance	Sales volume. Profit. Costs versus plan. Shareholder return. Associate/employee utilisation figures	Turnover £4m. Profit £400k. Return for shareholders. Days/month per person	
A growing base of satisfied customers	Sales/customer. Complaints/recommendations. Customer satisfaction	>£400k = 1 client. £100k-£400k =5 clients. £50k-£100k= 6 clients. <£50k=12 client	
A sufficient number of committed and competent people	No. of employed staff/associates. Gaps in competency matrix. Appraisal results. Perceptions of associates and staff	15 employed staff. 10 associates including 6 new by end of year	
Research projects properly completed and published	Proportion completed on time, in budget with customers satisfied. Number of publications per project	3 completed on time, in budget with satisfied customers	

Figure 3.8 CSF/core process reporting matrix

site/departmental level, Figure 3.9. This form of implementation should ensure the top down *and* bottom up approach to the deployment of policies and goals.

Deliverables

The deliverables then after one planning cycle of this process in a business will be:

1 An agreed framework for policy/goal deployment through the business.

2 Agreed mission statements for the business and, if required, for the business units/divisions.

3 Agreed critical success factors (CSFs) with ownership at top team level for the business and business units/divisions.

4 Agreed key performance indicators (KPIs) with targets throughout the business.

5 Agreed core business processes, with sponsorship at top team level.

6 A corporate CSF/business unit CSF matrix showing the impacts and the first 'whats/hows' deployment.

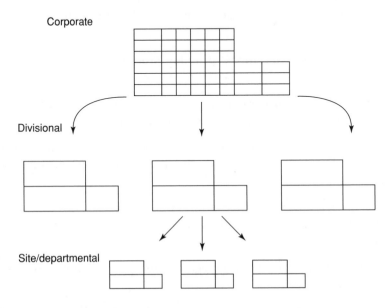

Figure 3.9 Deployment – what/how

7   A what/how (CSF/process) matrix approach for deploying the goals into the organization through process definition, understanding, and measured improvement at the business unit level.

8   Focused business improvement, linked back to the CSFs, with prioritized action plans and involvement of employees.

Strategic and operational planning

Changing the culture of an organization to incorporate a sustainable ethos of continuous improvement and responsive business planning will come about only as the result of a carefully planned and managed process. Clearly many factors are involved, including:

*   identifying strategic issues to be considered by the senior management team;

*   balancing the present needs of the business against the vital needs of the future;

*   concentrating finite resources on important things;

*   providing awareness of impending changes in the business environment in order to adapt more rapidly, and more appropriately.

Strategic planning is the continuous process by which any organization will describe its destination, assess barriers standing in the way of reaching that destination, and select approaches for dealing with those barriers and moving forward. Of course the real contributors to a successful strategic plan are the participants.

The strategic and operational planning process described in this chapter will:

*   Provide the senior management team with the means to manage the organization and strengths and weaknesses through the change process.

*   Allow the senior management team members to have a clear understanding and to achieve agreement on the strategic direction, including vision and mission.

*   Identify and document those factors critical to success (CSFs) in achieving the strategic direction and the means by which success will be measured (KPIs) and targeted.

*   Identify, document and encourage ownership of the core processes that drive the business.

*   Reach agreement on the priority processes for action by process improve-

ment teams, incorporating current initiatives into an overall, cohesive framework.

- Provide a framework for successfully deploying all goals and objectives through all organizational levels through a two-way 'catch-ball' process.

- Provide a mechanism by which goals and objectives are monitored, reviewed, and appropriate actions taken, at appropriate frequencies throughout the operational year

- Transfer the skills and knowledge necessary to sustain the process

The components outlined above will provide a means of effectively deploying a common vision and strategy throughout the organization. They will also allow for the incorporation of all change projects, as well as 'business as usual' activities, into a common framework which will form the basis of detailed operating plans.

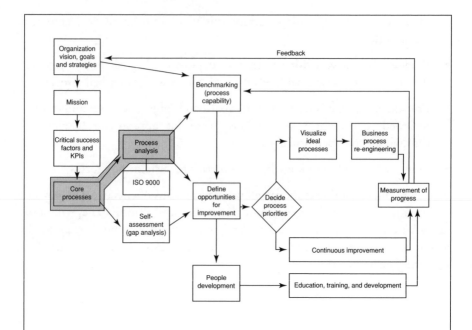

Quality and processes _____

Chapter 4

Quality and processes

Key points

The reputation enjoyed by any organization is built by quality, reliability, delivery and price, with quality perhaps the most important of these competitive weapons. Organizations 'delight' the customer by consistently meeting customer requirements, and achieve a reputation for 'excellence'.

In all organizations there exists a series of internal suppliers and customers. These form the so-called 'quality chains', the core of a total quality approach and organizational excellence.

Everything we do is a process, which is the transformation of a set of inputs into the desired outputs, and in every organization there are some core processes, which must be performed especially well if the vision and mission are to be achieved.

For example, the marketing processes establish the true requirements for the product or service and these must be communicated properly throughout the organization.

The involvement of all the people in an organization is a requirement of total organizational excellence, which means everyone working together at every interface to achieve excellence.

Quality and competitiveness

It does not matter which type of organization you work in – a hospital, a university, a bank, an insurance company, local government, an airline, a factory – competition is rife, competition for customers, for students, for patients, for resources, for funds. There are very few people around, in most types of organization, who remain to be convinced that quality is the most important of the competitive weapons. If you doubt that, just look at the way some organizations, even whole industries in certain countries, have used quality to take the heads off their competitors. And they are not only Japanese companies, there are British, American, French, German, Italian, Spanish, Swiss, Swedish ones, and organizations from other countries which have used quality strategically to win customers, steal business resources or funding, and be competitive. Moreover, attention to quality improves performance in reliability, delivery and price.

Thinking of Japanese companies, it is possible to remember a time when their reputation was anything but good. Not too long ago they were most famous for 'cheap trash'. They have clearly **learned** something. This has not as much to do with differences in national cultures as many people think it has. The Japanese culture, which is much older than most western cultures, has not changed significantly in less than 50 years.

One of the lessons many Japanese companies learned after World War II was to manage quality, and the other things on which we compete. They learned it from a handful of Americans, people like Joseph M. Juran and W. Edwards Deming who have since reached fame as 'gurus' of quality management.

For any organization, there are several aspects of **reputation** which are important:

i)   it is built upon the competitive elements: quality, reliability, delivery, and price, of which quality has become strategically the most important;

ii)   once an organization acquires a poor reputation for quality, it takes a very long time to change it;

iii) reputations, good or bad, can quickly become national reputations;

iv)   the management of the competitive weapons, such as quality, can be learned like any other skill, and used to eventually turn round a poor reputation.

Before anyone will buy the idea that quality is an important consideration, they would have to know what was meant by it.

What is quality?

'Is this a quality watch?' Pointing to my wrist I ask a class of students – undergraduates, postgraduates, experienced managers – it matters not. The answers vary:

'No, it's made in Japan.'
'No, it's cheap.'
'No, the face is scratched'
'How reliable is it?'
'I wouldn't wear it.'

My watch has been insulted all over the world – London, New York, Paris,

Sydney, Brussels, Amsterdam, Leeds! Very rarely am I told that the quality of the watch depends on what the wearer requires from a watch – a piece of jewellery to give an impression of wealth? A time piece which gives the required data, including the date, in digital form? An ability to perform at 50 metres under the sea? Clearly these requirements determine the quality.

Quality is often used to signify 'excellence' of a product or service – people talk about 'Rolls-Royce quality' and 'top quality'. In some engineering companies, the word may be used to indicate that a piece of metal conforms to certain physical dimension characteristics often set down in the form of a particularly 'tight' specification. In a hospital it might be used to indicate some sort of 'professionalism'. If we are to define quality in a way which is useful in its **management**, then we must recognize the need to include in the assessment of quality, the true requirements of the 'customer' – the needs and expectations.

Quality then is simply **meeting the customer requirements** and this has been expressed in many ways by other authors:

'fitness for purpose of use' – Juran.

'the totality of features and characteristics of a product or service that bear on its ability to satisfy stated or implied needs' – BS 4778, 1987 (ISO 8402, 1986) Quality Vocabulary: Part 1 International Terms.

'quality should be aimed at the needs of the consumer, present and future' – Deming.

'the total composite product and service characteristics of marketing, engineering, manufacture and maintenance through which the product and service in use will meet the expectation by the customer' – Feigenbaum.

'conformance to requirements' – Crosby.

There is another word that we should define properly – **reliability**. 'Why do you buy a BMW car?' 'Quality and reliability', comes back the answer. The two are used synonymously, often in a totally confused way. Clearly, part of the acceptability of a product or service will depend on its ability to function satisfactorily **over a period of time**, and it is this aspect of performance which is given the name **reliability**. It is the ability of the product or service to **continue** to meet the customer requirements. Reliability ranks with quality in importance, since it is a key factor in many purchasing decisions where alternatives are being considered. Many of the general management issues related to achieving product or service quality are also applicable to reliability.

It is important to realize that the 'meeting the customer requirements' definition of quality is not restrictive to the functional characteristics of products or services. Anyone with children knows that the quality of some of the products they purchase is more associated with **satisfaction in ownership** than some functional property. This is also true of many items, from antiques to certain items of clothing. The requirement for status symbols accounts for the sale of some executive cars, certain bank accounts and charge cards, and even hospital beds! The requirements are of paramount importance in the assessment of the quality of any product or service.

By **consistently** meeting customer requirements we can move to a different plane of satisfaction – **delighting the customer** and **customer loyalty**. There is no doubt that many organizations have so well addressed their capability to meet their customers' requirements time and time again, that this has created a reputation for 'excellence' and loyalty in customers.

Customers and the quality chains

The ability to meet the customer requirements is vital, not only between two separate organizations, but within the same organization.

The flight attendant, not thinking of quality problems, pulled back the curtain across the aisle and set off with a trolley full of breakfasts to feed the early morning travelling business people, on the short domestic flight into an international airport. Having stopped at the row of seats marked 1ABC, she passed the first tray onto the lap of the person sitting by the window. By the time the second tray had reached the person beside him, the first tray was on its way back to the flight attendant with a complaint that the bread roll and jam were missing. She calmly replaced it in her trolley and reached for another – which also had no roll and jam.

The calm exterior of the attendant began to evaporate as she discovered two more trays without a complete breakfast. Then she found a good one and, thankfully, passed it over. This search for complete breakfast trays continued down the aeroplane and caused the inevitable delays, so much so that several passengers did not receive their breakfasts until the plane had begun its descent. At the rear of the plane could be heard the mutterings of discontent. 'Aren't they slow with breakfast this morning?' 'What is the attendant doing with those trays?' 'We will have indigestion by the time we've landed.' It is interesting how the problem was perceived by many to be one of delivery or service. They could smell food but they weren't getting any of it, and they were getting really wound up! The flight attendant, who had suffered the embarrassment of being the purveyor of defective product and service, returned to the curtain and almost ripped it from the hooks. She was heard to say through clenched teeth, 'What a bloody mess!'

A problem of quality? Yes, of course, requirements not being met, but where? The passengers or customers suffered from it on the aircraft, but down in the bowels of the organization there was a person whose job it was to assemble the breakfast trays. On this day the system broke down – perhaps he ran out of bread rolls, perhaps he was called away to refuel the aircraft (it was a small airport!), perhaps he didn't know or understand, perhaps he didn't care.

Three hundred miles away in a chemical factory ... 'What the hell is quality control doing? We've just sent 15 000 litres of lawn weedkiller to CIC and there it is back at our gate – they've returned it as out of spec.' This was followed by an avalanche of verbal abuse, which will not be repeated here, but poured all over the shrinking quality control manager as he backed through his office door, followed by a red faced technical director advancing menacingly from behind the bottles of sulphuric acid racked across the adjoining laboratory.

'Yes, what is QC doing?' thought the production manager, who was behind a door two offices along the corridor, but could hear the torrent of language now being used to beat the QC man into an admission of guilt. He knew the poor devil couldn't possibly do anything about the rubbish that had been produced except test it, but why should he volunteer for the unpleasant and embarrassing ritual now being experienced by his colleague – for the second time this month. No wonder the QC manager had been studying the appointments pages of the *Telegraph* on Thursday – what a job!

Do you recognize these two situations? Do they not happen every day of the week – possibly every minute somewhere in manufacturing or the service industries? Is it any different in banking, insurance, the health service?

The inquisition of checkers and testers – the last bastion of desperate systems which try in vain to catch mistakes, stop defectives, hold lousy materials, before they reach the external customer – and woe betide the idiot who lets it pass through!

Two everyday incidents, but why are events like these so common? The answer is the acceptance of one thing – **failure** – to do it right the first time every time at every stage.

Why do we accept failure in the production of artefacts, the provision of a service, or even the transfer of information? In many walks of life, we do not accept it. We do not say, 'Well, the nurse is bound to drop the odd baby in a thousand – it's just going to happen.' We do not accept that!

There exists in each department, each office, even each household, a series of suppliers and customers. The typist is a supplier to her supervisor – is she meeting his or her requirements? Is error free typing set out as wanted, when wanted? If so, then we have a quality typing service. Does the flight attendant receive from the supplier in the airline the correct food trays in the right quantity?

Throughout and beyond all organizations, whether they be manufacturing concerns, banks, retail stores, universities, hospitals or hotels, there is a series of **quality chains** of customer and suppliers (Figure 4.1) which may be broken at any point by one person or one piece of equipment not meeting the requirements of the customer, internal or external. The interesting point is that this failure usually finds its way to the interface between the organization and its outside customers, and the people who operate at that interface – like the flight attendant – usually experience the ramifications. The concept of internal and external customers/suppliers forms the **core** of the total quality approach.

A great deal is written and spoken about employee motivation as a separate issue. In fact the key to motivation and quality is for everyone in the organization to have well-defined customers – an expansion of the word, beyond the

Outside organization

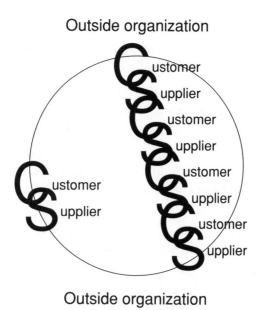

Outside organization

Figure 4.1 The quality chains

outsider that actually purchases or uses the ultimate product or service, to anyone to whom an individual gives a part, a service, information; in other words the results of his or her work.

Quality has to be managed – it will not just happen. Clearly it must involve everyone in the process and be applied throughout the organization. Many people in the support functions of organizations never see, experience, or touch the products or services that their organizations purchase or provide, but they do handle or produce things like purchase orders or invoices. If every fourth invoice carries at least one error, what image of quality is transmitted?

Failure to meet the requirements in any part of a quality chain have a way of multiplying, and failure in one part of the system creates problems elsewhere, leading to yet more failure, more problems and so on. The price of quality is the continual examination of the requirements and our ability to meet them. This alone will lead to a 'continuing improvement' philosophy. The benefits of making sure the requirements are met at every stage, every time, are truly enormous in terms of increased competitiveness and market share, reduced costs, improved productivity and delivery performance, and the elimination of waste.

Meeting the requirements

If quality is meeting the customer requirements then this has wide implications. The requirements may include availability, delivery, reliability, maintainability and cost effectiveness, amongst many other features. The first item on the list of things to do is find out what the requirements are. If we are dealing with a customer/supplier relationship crossing two organizations, then the supplier must establish a 'marketing' activity charged with this task.

The marketers must, of course, understand not only the needs of the customer, but also the ability of their own organization to meet the demands. If my customer places a requirement on me to run 1500 metres in four minutes, then I know I am unable to meet this demand, unless something is done to improve my running performance. Of course, I may never be able to achieve this requirement.

Within organizations, between internal customers and suppliers, the transfer of information regarding requirements is frequently poor to totally absent. How many executives really bother to find out what their customers' – their secretaries' – requirements are? Can their handwriting be read, do they leave clear instructions, do the secretaries always know where the boss is? Equally, do the secretaries establish what their bosses need – error free typing, clear messages, a tidy office? The internal supplier/customer relationships are often the most difficult to manage in terms of establishing the requirements. To achieve quality throughout an organization, each person in the quality chain must interrogate every interface as follows:

Customers

- Who are my immediate customers?

- What are their true requirements?

- How do or can I find out what the requirements are?

- How can I measure my ability to meet the requirements?
  (If not then what must change to improve the capability?)

- Do I continually meet the requirements?
  (If not then what prevents this from happening, when the capability exists?)

- How do I monitor changes in the requirements?

Suppliers

- Who are my immediate suppliers?

- What are my true requirements?

- How do I communicate my requirements?

- Do my suppliers have the capability to measure and meet the requirements?

- How do I inform them of changes in the requirements?

The measurement of capability is extremely important if the quality chains are to be formed within and without an organization. Each person in the organization must also realize that the supplier's needs and expectations must be respected if the requirements are to be fully satisfied.

Design and conformance

To understand how quality may be built into a product or service, at any stage, it is necessary to examine the two distinct, but interrelated aspects of quality:

- Quality of design

- Quality of conformance to design

Quality of design

Design, like quality, permeates many areas of an organization at both the strategic and operational levels. Quality of design is a measure of how well the product or service is designed to achieve the agreed requirements. The beautifully presented gourmet meal will not necessarily please the recipient if he or she is travelling on the highway and stopped for a quick bite to eat. The most important feature of the design, with regard to achieving quality, is the specification. Specifications must exist at the internal and external supplier/customer interfaces to pursue company-wide quality. For example, the company lawyer asked to draw up a contract by the sales manager requires a specification as to its content:

- Is it a sales, processing or consulting type of contract?

- Who are the contracting parties?

- In which countries are the parties located?

- What are the products involved (if any)?

- What is the volume?

- What are the financial (e.g. price, escalation) aspects?

The financial controller must issue a specification of the information he or she needs, and when, to ensure that foreign exchange fluctuations do not cripple the company's finances. The business of sitting down and agreeing a specification at every interface will clarify the true requirements and capabilities. It is the vital first stage for a successful total quality effort.

While functions like graphics or styling may be largely the preserve of design professionals, design decisions about new services and products are likely to involve just as many non-designers – from R&D to finance to marketing. It has been widely recognized for many years that success in any organization is more likely to occur when design, marketing and production/operations are well interfaced and co-ordinated. An organization's design capability and the quality of product/service are inextricably linked. Quality must be 'designed in' to the new product and success and competitive advantage can be sustained only when the design activity is well understood and integrated across the organization's different functions.

Quality of conformance to design

There must be a corporate understanding of the organization's position in the market place. It is not sufficient that marketing specifies the product or service, 'because that is what the customer wants'. There must be an agreement that the operating departments can achieve that requirement. Should they be incapable of doing so, then one of two things must happen, either the organization finds a different position in the market place or substantially changes the operational facilities. Quality of conformance to design is the

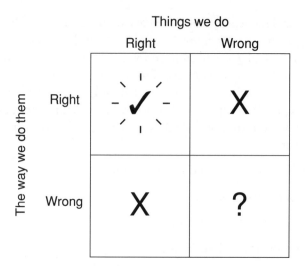

Figure 4.2 How much time is spent doing the right things right?

extent to which the product or service achieves the quality of design. What the customer actually receives should conform to the design, and operating costs are tied firmly to the level of conformance achieved. Quality cannot be inspected into products or services, the customer satisfaction must be designed into the whole system. The conformance check then makes sure that things go according to plan.

A high level of inspection or checking at the end is often indicative of attempts to inspect in quality. This may well result in spiralling costs and decreasing viability. The area of conformance to design is concerned largely with the performance of the actual operations. It may be salutary for organizations to use the simple matrix of Figure 4.2 to assess how much time they spend doing the right things right. A lot of people often through no fault of their own spend a good proportion of the available time doing the right things wrong. There are people (and organizations) who spend time doing the wrong things very well, and even those who occupy themselves doing the wrong things wrong, which can be very confusing!

Managing processes

Every day two men who work in a certain factory scrutinize together the results of the examination of the previous day's production, and commence the ritual battle over whether the material is suitable for despatch to the customer. One is called the production manager, the other the quality control manager. They argue and debate the evidence before them, the rights and wrongs of the specification, and each tries to convince the other of the validity of his argument. Sometimes they nearly break into fighting.

This ritual is associated with trying to answer the question:

'Have we done the job correctly?'

'correctly' being a flexible word depending on the interpretation given to the specification on that particular day. This is not **control**, it is **detection**, post-production, wasteful detection of bad product before it hits the customer. There is still a belief in some quarters that to achieve quality we must check, test, inspect or measure – the ritual pouring on of quality at the end of the process. This is nonsense, but it is frequently practised. In the office one finds staff checking other people's work before it goes out, validating computer input data, checking invoices, typing, etc. There is also quite a lot of looking for things, chasing why things are late, apologizing to customers for lateness, and so on – waste, waste, waste.

To get away from the natural tendency to rush into the detection mode, it is necessary to ask different questions in the first place. We should not ask whether the job has been done correctly, we should ask first:

'Are we capable of doing the job correctly?'

This has wide implications and much of this book is devoted to the various activities which are necessary to ensure that the answer is yes. However, we should realize straight away that such an answer will only be obtained using satisfactory methods, materials, equipment, skills and instruction, and a satisfactory 'process'.

What is a process?

As we have seen, quality chains can be traced right through the business or service processes used by any organization. A process is the transformation of a set of inputs, which can include actions, methods and operations, into desired outputs which satisfy the customer needs and expectations, in the form of products, information, services or – generally – results. Everything we do is a process, so in each area or function of an organization there will be many processes taking place. For example, a finance department may be involved in budgeting processes, accounting processes, salary and wage processes, costing processes, etc. Each process in each department or functional area can be analysed by an examination of the inputs and outputs. This will determine some of the actions necessary to improve performance.

The output from a process is that which is transferred to somewhere or to someone – the **customer**. Clearly to produce an output which meets the

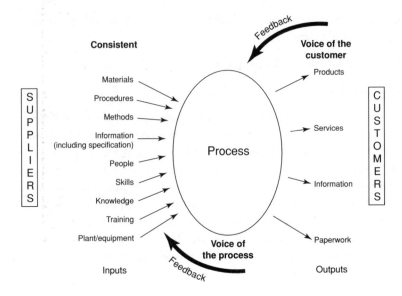

Figure 4.3 A process

requirements of the customer, it is necessary to define, monitor and control the inputs to the process, which in turn may be supplied as output from an earlier process. At every supplier–customer interface then there resides a transformation process (Figure 4.3), and every single task throughout an organization must be viewed as a process in this way. Once we have established that our process is capable of meeting the requirements, we can address the next question,

'Do we continue to do the job correctly?'

which brings a requirement to monitor the process and the controls on it. If we now re-examine the first question: 'Have we done the job correctly?' we can see that, if we have been able to answer the other two questions with a yes, we **must** have done the job correctly – any other outcome would be illogical. By asking the questions in the right order, we have moved the need to ask the 'inspection' question and replaced a strategy of **detection** with one of **prevention**. This concentrates all the attention on the front end of any process – the inputs – and changes the emphasis to making sure the inputs are capable of meeting the requirements of the process. This is a managerial responsibility.

These ideas apply to every transformation process, which must be subject to the same scrutiny of the methods, the people, skills, equipment and so on to make sure they are correct for the job. A person giving a lecture, whose overhead projector equipment will not focus correctly, or whose teaching materials are not appropriate, will soon discover how difficult it is to provide a lecture which meets the requirements of the audience.

As we saw in the previous chapter, in every organization there are some very large processes – groups of smaller processes called '**core business processes**'. These are activities which the organization must carry out especially well if its mission and objectives are to be achieved. This is crucial if the management of processes is to be integrated into the strategy for the organization.

The **control** of quality clearly can only take place at the point of operation or production – where the letter is typed, the sales call made, the patient admitted, or the chemical manufactured. The act of **inspection is not quality control**. When the answer to, 'Have we done the job correctly?' is given indirectly by answering the questions of capability and control, then we have **assured** quality and the activity of checking becomes one of **quality assurance** – making sure that the product or service represents the output from an effective **system** to ensure capability and control. It is frequently found that organizational barriers between departmental empires encourage the development of testing and checking of services or products in a vacuum, without interaction with other departments.

Excellence starts with 'marketing'

The author has been asked on more than one occasion if all this applies to marketing? The answer to the question is not remarkable – it starts there!

The 'marketing' function of an organization must take the lead in establishing the true requirements for the product or service. Having determined the need, marketing should define the market sector and demand. This will determine product or service features such as the grade, price, quality, timing, etc. For example, a major hotel chain, before opening a new hotel or refurbishing an old one, will need to consider its location and accessibility, before deciding whether it will be predominantly a budget, first class, business or family hotel.

Marketing will also need to establish customer requirements by reviewing the market needs, particularly in terms of unclear or unstated expectations or preconceived ideas held by customers. Marketing is responsible for determining the key characteristics which determine the suitability of the product or service in the eyes of the customer. This may, of course, involve the use of market research techniques, data gathering, and analysis of customer complaints. If possible, quasi-quantitative methods should be employed giving proxy variables which can be used to grade the characteristics in importance, and decide in which areas superiority over competitors exists. It is often useful to compare these findings with internal perceptions of quality.

Excellent communication between customers and suppliers is the key to total organizational excellence. This will eradicate the 'demanding nuisance/idiot' view of customers, which pervades some organizations. Poor communications often occur in the supply chain between organizations, when neither party realizes how poor they are. Feedback from both customers and suppliers needs to be improved, where dissatisfied customers and suppliers do not communicate their problems. In such cases non-conformance of purchased products or services is often due to the customer's inability to communicate their requirements clearly. If these ideas are also used within an organization, then the internal supplier/customer interfaces will operate much more smoothly.

All the efforts devoted to finding the nature and timing of the demand will be pointless if marketing fails to communicate the requirements promptly, clearly, and accurately to the remainder of the organization. The marketing function should be capable of supplying the organization with a formal statement or outline of the requirements for each product or service. This constitutes a preliminary set of **specifications** which can be used as the basis for service or product design.

Marketing must also establish systems for feedback of customer information and reaction, which should be designed on a continuous monitoring basis. Any information pertinent to the product or service should be collected and collated, interpreted, analysed, and communicated to improve the response to customer experience and expectations. These same principles must also be applied inside the organization for continuous improvement at every transformation process interface to be achieved. If one department has problems recruiting the correct sort of staff, and 'HR' have not established mechanisms for gathering, analysing, and responding to information on new employees, then frustration and conflict will replace communication and co-operation.

Excellence in all functions

For an organization to be truly excellent, each part of it must work properly together. Each part, each activity, each person in the organization affects and is in turn affected by others. Errors have a way of multiplying and failure to meet the requirements in one part or area creates problems elsewhere, leading to yet more errors, yet more problems and so on. The benefits of getting it right first time everywhere are enormous.

Everyone experiences – almost accepts – problems in working life. This causes people to spend a large part of their time on useless activities, correcting errors, looking for things, finding out why things are late, checking suspect information, rectifying and reworking, apologizing to customers for mistakes, poor quality and lateness. The list is endless and it is estimated that about one-third of our efforts are wasted in this way. In the service sector it can be much higher.

Quality, the way we have defined it as meeting the customer requirements, gives people in different functions of an organization a common language for improvement. It enables all the people, with different abilities and priorities, to communicate readily with one another, in pursuit of a common goal. When business and industry was local, the craftsman could manage more or less on his own. Business is now so complex and employs so many different specialist skills that everyone has to rely on the activities of others in doing their jobs.

Some of the most exciting applications of process management have mate-rialized from departments which could see little relevance when first intro-duced to the concepts. Following training, many examples from different departments of organizations show the use of the techniques. Sales staff can monitor and increase successful sales calls, office staff have used the methods to prevent errors in word-processing and improve inputting to computers, customer service people have monitored and reduced complaints, distribution have controlled lateness and disruption in deliveries.

It is worthy of mention that the first point of contact for some outside cus-tomers is the telephone operator, the security people at the gate, or the person in reception. Equally, the paperwork and support services associated with the product, such as invoices and sales literature, must match the needs of the cus-tomer. Clearly organizational excellence cannot be restricted to the 'produc-tion' or operational areas without losing great opportunities to gain maximum benefit.

Management which rely heavily on exhortation of the workforce to 'do the right job right the first time', or 'accept that quality is your responsibility', will not only fail to achieve excellence but will create division and conflict. These calls for improvement infer that faults are caused only by the workforce and that problems are departmental when, in fact, the opposite is true – most problems are inter-departmental. The involvement of all members of an organization is a requirement of 'organization-wide improvement'. It must involve everyone working together at every interface to achieve perfection, and that can only happen if the processes that people work in are understood, well managed and continuously improved.

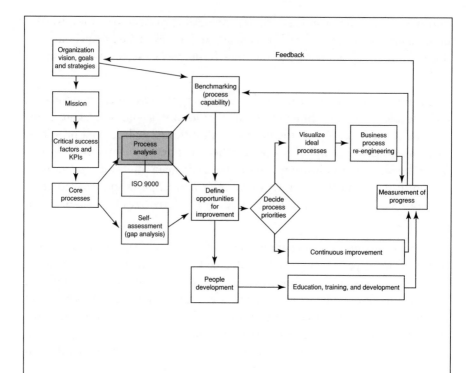

Process analysis _____

Chapter 5

Process analysis _____

Key points

The Process Classification Framework provides a high-level generic model for business processes found in both commercial and public sector organizations. Its aim is to help managers identify the main processes and sub-processes in their own organizations, and lay the foundation for systems and procedures.

Process modelling, mapping and flowcharting are methods of describing a process in pictures, using symbols to improve knowledge of the process and as an aid to develop the team of people involved.

Flowcharts are of four basic types: person, material, equipment, and information. These charts and diagrams may be examined using questioning techniques to determine: purpose, place, sequence, people, and method, to eliminate, combine, rearrange, or simplify process steps. Flowcharts document processes and are useful in troubleshooting and in process improvement.

Ten steps for process management and improvement are provided.

The Process Classification Framework

The Process Classification Framework was developed and copyrighted by the American Productivity and Quality Center (APQC) International Benchmarking Clearinghouse, with the assistance of several major international corporations, and in close partnership with Arthur Andersen & Co.

The intent was to create a high-level generic enterprise model that encourages businesses and other organizations to see their activities from a cross-industry, process viewpoint rather than from a narrow functional viewpoint.

The Process Classification Framework supplies a generic view of business processes often found in multiple industries and sectors – manufacturing and service companies, health care, government, education, and others. Many organizations now seek to understand their inner workings from a horizontal, process viewpoint, rather than from a vertical, functional viewpoint (Figure 5.1). How can they, for example, differentiate the sales *process* from the existing sales *department*?

The Process Classification Framework seeks to represent major processes

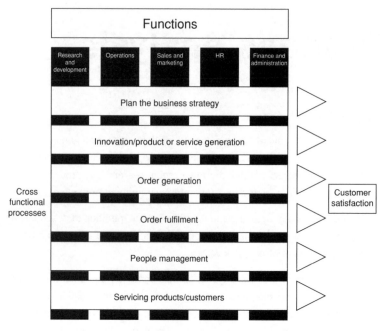

Figure 5.1   Cross-functional approach to managing core business processes

and sub-processes, not functions, through its structure (Figure 5.2) and vocabulary (Table 5.1). The framework does not list all processes within any specific organization. Likewise, not every process listed in the framework is present in every organization.

About the framework

The Process Classification Framework was originally envisioned as a 'taxonomy' of business processes during the 1991 design of the American Productivity & Quality Center's International Benchmarking Clearinghouse. That design process involved more than 80 organizations with a strong interest in advancing the use of benchmarking in the USA and around the world.

The founding members of the clearinghouse were convinced that a common vocabulary, not tied to any specific industry, was necessary to classify information by process and to help companies transcend the limitations of 'insider' terminology. A small team representing both industry and the center held the initial design meetings in early 1992. The center published the first version of the framework later that year.

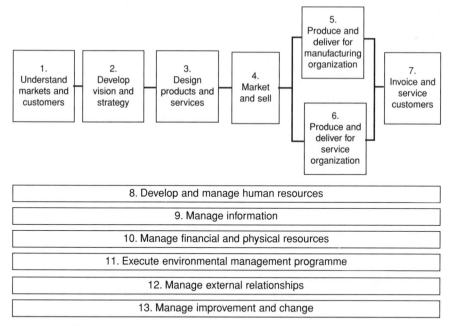

Figure 5.2   Process Classification Framework: overview

Continuing dialogue with clearinghouse members has shown that the Process Classification Framework can be a useful tool in understanding and mapping business processes. In particular, a number of organizations have used the framework to classify both internal and external information for the purpose of cross-functional and cross-divisional communication.

The Process Classification Framework is an evolving document and the center will continue to enhance and improve it on a regular basis. To that end, the center welcomes your comments, suggestions for improvement, and any insights you gain from applying it within your organization. The center would like to see the Process Classification Framework receive wide distribution, discussion, and use. Therefore, it grants permission for copying the framework, as long as acknowledgement is made to the American Productivity & Quality Center.*

*Please direct your comments, suggestions, and questions to:
APQC International Benchmarking Clearinghouse Information Services Dept.
123 North Post Oak Lane, 3rd Floor, Houston, Texas 77024.
713-681-4020 (phone)
713-681-8578 (fax)
Internet: apqcinfo@apqc.org
For updates, visit the Web site at http://www.apqc.org

Table 5.1   Process classification framework – vocabulary

OPERATING PROCESSES	

1 Understand markets and customers

1.1 Determine customer needs and wants

  1.1.1 Conduct qualitative assessments

    1.1.1.1 Conduct customer interviews

    1.1.1.2 Conduct focus groups

  1.1.2 Conduct quantitative assessments

    1.1.2.1 Develop and implement surveys

  1.1.3 Predict customer purchasing behaviour

1.2 Measure customer satisfaction

  1.2.1 Monitor satisfaction with products and services

  1.2.2 Monitor satisfaction with complaint resolution

  1.2.3 Monitor satisfaction with communication

1.3 Monitor changes in market or customer expectations

  1.3.1 Determine weaknesses of product/service offerings

  1.3.2 Identify innovations that meet customer needs

  1.3.3 Determine customer reactions to competitive offerings.

2 Develop vision and strategy

2.1 Monitor the external environment

  2.1.1 Analyse and understand competition

  2.1.2 Identify economic trends

  2.1.3 Identify political and regulatory issues

  2.1.4 Assess technology innovations

  2.1.5 Understand demographics

  2.1.6 Identify social and cultural changes

  2.1.7 Understand ecological concerns

2.2 Define the business concept and organisational strategy

  2.2.1 Select relevant markets

  2.2.2 Develop long-term vision

  2.2.3 Formulate business unit strategy

  2.2.4 Develop overall mission statement

2.3 Design the organizational structure and relationships between organizational units

2.4 Develop and set organizational goals.

3 Design products and services

3.1 Develop new product/service concept and plans

  3.1.1 Translate customer wants and needs into product and/or service requirements

  3.1.2 Plan and deploy quality targets

  3.1.3 Plan and deploy cost targets

  3.1.4 Develop product life cycle and development timing targets

  3.1.5 Develop and integrate leading technology into product/service concept

3.2 Design, build, and evaluate prototype products and services

  3.2.1 Develop product/service specifications

  3.2.2 Conduct concurrent engineering

  3.2.3 Implement value engineering

  3.2.4 Document design specifications

  3.2.5 Develop prototypes

  3.2.6 Apply for patents

3.3 Refine existing products/services

  3.3.1 Develop product/service enhancements

  3.3.2 Eliminate quality/reliability problems

  3.3.3 Eliminate outdated products/services

3.4 Test effectiveness of new or revised products or services

3.5 Prepare for production

3.5.1 Develop and test prototype production process

3.5.2 Design and obtain necessary materials and equipment

3.5.3 Install and verify process or methodology

3.6 Manage the product/service development process

4 Market and sell

4.1 Market products or services to relevant customer segments

4.1.1 Develop pricing strategy

4.1.2 Develop advertising strategy

4.1.3 Develop marketing messages to communicate benefits

4.1.4 Estimate advertising resource and capital requirements

4.1.5 Identify specific target customers and their needs

4.1.6 Develop sales forecast

4.1.7 Sell products and services

4.1.8 Negotiate terms

4.2 Process customer orders

4.2.1 Accept orders from customers

4.2.2 Enter orders into production and delivery process

5 Produce and deliver for manufacturing oriented organizations

5.1 Plan for and acquire necessary resources

5.1.1 Select and certify suppliers

5.1.2 Purchase capital goods

5.1.3 Purchase materials and supplies

5.1.4 Acquire appropriate technology

5.2 Convert resources or inputs into products

5.2.1 Develop and adjust production delivery process (for existing process)

5.2.2 Schedule production

5.2.3 Move materials and resources

5.2.4 Make product

5.2.5 Package product

5.2.6 Warehouse or store product

5.2.7 Stage products for delivery

5.3 Deliver products

5.3.1 Arrange product shipment

5.3.2 Deliver products to customers

5.3.3 Install product

5.3.4 Confirm specific service requirements for individual customers

5.3.5 Identify and schedule resources to meet service requirements

5.3.6 Provide the service to specific customers

5.4 Manage production and delivery process

5.4.1 Document and monitor order status

5.4.2 Manage inventories

5.4.3 Assure product quality

5.4.4 Schedule and perform maintenance

5.4.5 Monitor environmental constraints

6 Produce and deliver for service oriented organizations

6.1 Plan for and acquire necessary resources

6.1.1 Select and certify suppliers

6.1.2 Purchase materials and supplies

6.1.3 Acquire appropriate technology

6.2 Develop human resource skills

6.2.1 Define skill requirements

6.2.2 Identify and implement training

6.2.3 Monitor and manage skill development

6.3 Deliver service to the customer

6.3.1 Confirm specific service requirements for individual customer

6.3.2 Identify and schedule resources to meet service requirements

6.3.3 Provide the service to specific customers

6.4 Ensure quality of service

7 Invoice and service customers

7.1 Bill the customer

   7.1.1 Develop, deliver and maintain customer billing

   7.1.2 Invoice the customer

   7.1.3 Respond to billing inquiries

7.2 Provide after-sales service

   7.2.1 Provide post-sales service

   7.2.2 Handle warranties and claims

7.3 Respond to customer inquiries

   7.3.1 Respond to information requests

   7.3.2 Manage customer complaints

MANAGEMENT & SUPPORT PROCESSES

8 Develop and manage human resources

8.1 Create and manage human resource strategies

   8.1.1 Identify organizational strategic demands

   8.1.2 Determine human resource costs

   8.1.3 Define human resource requirements

   8.1.4 Define human resources' organizational role

8.2 Cascade strategy to work level

   8.2.1 Analyse, design, or redesign work

   8.2.2 Define and align work outputs and metrics

   8.2.3 Define work competencies

8.3 Manage deployment of personnel

   8.3.1 Plan and forecast workforce requirements

   8.3.2 Develop succession and career plans

   8.3.3 Recruit, select and hire employees

   8.3.4 Create and deploy teams

   8.3.5 Relocate employees

   8.3.6 Restructure and rightsize workforce

   8.3.7 Manage employee retirement

   8.3.8 Provide outplacement support

8.4 Develop and train employees

   8.4.1 Align employee and organization development needs

   8.4.2 Develop and manage training programmes

   8.4.3 Develop and manage employee orientation programmes

   8.4.4 Develop functional/process competencies

   8.4.5 Develop management/leadership competencies

   8.4.6 Develop team competencies

8.5 Manage employee performance, reward and recognition

   8.5.1 Define performance measures

   8.5.2 Develop performance management approaches/feedback

   8.5.3 Manage team performance

   8.5.4 Evaluate work for market value and internal equity

   8.5.5 Develop and manage base and variable compensation

   8.5.6 Manage reward and recognition programmes

8.6 Ensure employee well-being and satisfaction

   8.6.1 Manage employee satisfaction

   8.6.2 Develop work and family support systems

   8.6.3 Manage and administer employee benefits

   8.6.4 Manage workplace health and safety

   8.6.5 Manage internal communications

   8.6.6 Manage and support workforce diversity

8.7 Ensure employee involvement

8.8 Manage labour-management relationships

   8.8.1 Manage collective bargaining process

   8.8.2 Manage labour-management partnerships

8.9 Develop Human Resource Information Systems (HRIS)

9 Manage information resources

9.1 Plan for information resource management
   9.1.1 Derive requirements from business strategies
   9.1.2 Define enterprise system architectures
   9.1.3 Plan and forecast information technologies/methodologies
   9.1.4 Establish enterprise data standards
   9.1.5 Establish quality standards and controls

9.2 Develop and deploy enterprise support systems
   9.2.1 Conduct specific needs assessments
   9.2.2 Select information technologies
   9.2.3 Define data life cycles
   9.2.4 Develop enterprise support systems
   9.2.5 Test, evaluate, and deploy enterprise support systems

9.3 Implement systems security and controls
   9.3.1 Establish systems security strategies and levels
   9.3.2 Test, evaluate, and deploy systems security and controls

9.4 Manage information storage and retrieval
   9.4.1 Establish information repositories (databases)
   9.4.2 Acquire and collect information
   9.4.3 Store information
   9.4.4 Modify and update information
   9.4.5 Enable retrieval of information
   9.4.6 Delete information

9.5 Manage facilities and network operation
   9.5.1 Manage centralized facilities
   9.5.2 Manage distributed facilities

9.5.3 Manage network operations

9.6 Manage information services
   9.6.1 Manage libraries and information centres
   9.6.2 Manage business records and documents

9.7 Facilitate information sharing and communication
   9.7.1 Manage external communications systems
   9.7.2 Manage internal communications systems
   9.7.3 Prepare and distribute publications

9.8 Evaluate and audit information quality

10 Manage financial and physical resources

10.1 Manage financial resources
   10.1.1 Develop budgets
   10.1.2 Manage resource allocation
   10.1.3 Design capital structure
   10.1.4 Manage cash flow
   10.1.5 Manage financial risk

10.2 Process finance and accounting transactions
   10.2.1 Process accounts payable
   10.2.2 Process payroll
   10.2.3 Process accounts receivable, credit and collections
   10.2.4 Close the books
   10.2.5 Process benefits and retiree information
   10.2.6 Manage travel and entertainment expenses

10.3 Report information
   10.3.1 Provide external financial information
   10.3.2 Provide internal financial information

10.4 Conduct internal audits

10.5 Manage the tax function
   10.5.1 Ensure tax compliance
   10.5.2 Plan tax strategy
   10.5.3 Employ effective technology

10.5.4 Manage tax controversies
10.5.5 Communicate tax issues to
management
10.5.6 Manage tax administration
10.6 Manage physical resources
10.6.1 Manage capital planning
10.6.2 Acquire and redeploy fixed
assets
10.6.3 Manage facilities
10.6.4 Manage physical risk

**11 Execute environmental management
programme**
11.1 Formulate environmental
management strategy
11.2 Ensure compliance with regulations
11.3 Train and educate employees
11.4 Implement pollution prevention
programme
11.5 Manage remediation efforts
11.6 Implement emergency response
programmes
11.7 Manage government agency and
public relations
11.8 Manage acquisition/divestiture
environmental issues
11.9 Develop and manage environmental
information system
11.10 Monitor environmental
management

12 Manage external relationships
12.1 Communicate with shareholders
12.2 Manage government relationships
12.3 Build lender relationships
12.4 Develop public relations programme
12.5 Interface with board of directors

12.6 Develop community relations
12.7 Manage legal and ethical issues

13 Manage improvement and change
13.1 Measure organizational performance
13.1.1 Create measurement systems
13.1.2 Measure product and
service quality
13.1.3 Measure cost of quality
13.1.4 Measure costs
13.1.5 Measure cycle time
13.1.6 Measure productivity
13.2 Conduct quality assessments
13.2.1 Conduct quality assessments
based on external criteria
13.2.2 Conduct quality assessments
based on internal criteria
13.3 Benchmark performance
13.3.1 Develop benchmarking
capabilities
13.3.2 Conduct process
benchmarking
13.3.3 Conduct competitive
benchmarking
13.4 Improve processes and systems
13.4.1 Create commitment for improvement
13.4.2 Implement continuous process
improvement
13.4.3 Reengineer business
processes and systems
13.4.4 Manage transition to change
13.5 Implement TQM
13.5.1 Create commitment for TQM
13.5.2 Design and implement TQM
systems
13.5.3 Manage TQM life cycle

Process modelling

Between 1974 and 1981 the US Airforce adopted 'Integration Definition Function Modelling' (IDEFO), as part of its Integrated Computer-Aided Manufacturing (ICAM) architecture. The IDEFO modelling language, now described in a Federal Information Processing Standards Publication (FIPS

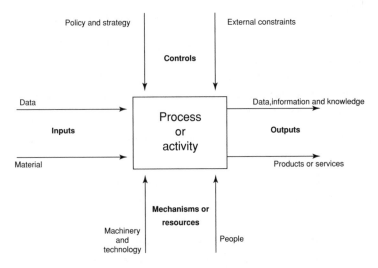

Figure 5.3 IDEFO process language. Federal Information Processing Standard Publication 183 (December 1993) National Institute of Standards and Technology (NIST)

PUBS) provides a useful structured graphical framework for describing and improving business processes. The associated 'Integration Definition for Information Modelling' (IDEFIX) language allows the development of a logical model of data associated with processes, such as measurement.

These techniques are widely used in business process re-engineering (BPR) and business process improvement (BPI) projects, and to integrate process information. A range of specialist software (including Windows/PC based) is also available to support the applications. IDEFO may be used to model a wide variety of new and existing processes, define the requirements, and design an implementation to meet the requirements.

An IDEFO model consists of a hierarchical series of diagrams, text and glossary, cross-referenced to each other through boxes (process components) and arrows (data and objects). The method is expressive and comprehensive and is capable of representing a wide variety of business, service and manufacturing processes. The relatively simple language allows coherent, rigorous and precise process expression, and promotes consistency (Figure 5.3).

For a full description of the IDEFO methodology it is necessary to consult the FIPS PUBS standard (Federal Information Processing Standard Publication 183 (December 1993), National Institute of Standards and Technology (NIST)). It should be possible, however, from the simple description given here, to begin process modelling (or mapping) using the technique.

Processes can be any combination of things, including people, information, software, equipment, systems, products or materials. The IDEFO model describes what a process does, what controls it, what things it works on, what means it uses to perform its functions, and what it produces. The combined graphics and text are comprised of:

Boxes   which provide a description of what happens in the form of an active verb or verb phrase.

Arrows   which convey data or objects related to the processes to be performed (they do not represent flow or sequence as in the traditional process flow model)

Each side of the process box has a standard meaning in terms of box/arrow relationships. Arrows on the left side of the box are **inputs**, which are transformed or consumed by the process to produce **output** arrows on the right side. Arrows entering the top of the box are **controls** which specify the conditions required for the process to generate the correct outputs. Arrows connected to the bottom of the box represent '**mechanisms**' or **resources**.

Using these relationships, process diagrams are broken down or decomposed into more detailed diagrams, the top level diagram providing a description of the highest level process. This is followed by a series of child diagrams providing details of the sub-processes (see Figure 5.4).

Each process model has a top-level diagram on which the process is represented by a single box with its surrounding arrows (e.g. Figure 5.5). Each sub-process is modelled individually by a box, with parent boxes detailed by child diagrams at the next lower level (e.g. Figures 5.6. and 5.7).

Text and glossary

An IDEFO diagram may have associated structured text to give an overview of the process model. This may also be used to highlight features, flows, and inter-box connections and to clarify significant patterns. A glossary may be used to define acronyms, key words and phrases used in the diagrams.

Arrows

Arrows on high-level IDEFO diagrams represent data or objects as constraints. Only at low levels of detail can arrows represent flow or sequence. These high-level arrows may usefully be thought of as pipelines or conduits with general labels. On lower level diagrams, arrows have more specific labels. An arrow may branch, fork or join indicating that the same kind of data or object may be needed or produced by more than one process or sub-process.

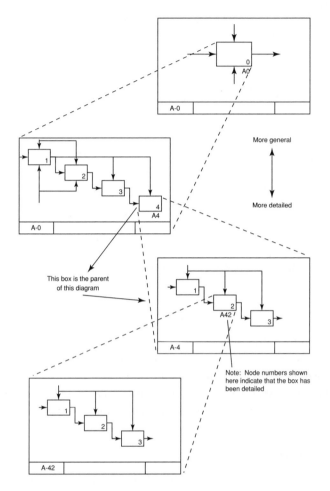

Figure 5.4   Decomposition structure – sub-processes

IDEFO process modelling, improvement and teamwork

The IDEFO methodology includes procedures for developing and critiquing process models by a group or team of people. The creation of an IDEFO process model provides a disciplined teamwork procedure for process under-standing and improvement. As the group works on the process following the discipline, the diagrams are changed to reflect corrections and improvements. More detail can be added by creating more diagrams, which in turn can be reviewed and altered. The final model represents an agreement on the process for a given purpose and from a given viewpoint, and can be the basis of new process or system improvement projects.

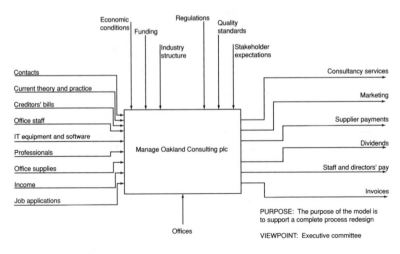

Figure 5.5 A top-level process diagram

IDEFIX

This is used to produce structural graphical information models for processes, which may support the management of data, the integration of information systems, and the building of computer databases. It is described in detail in the FIPS PUB 184 (December 1993, NIST). Its use is facilitated by the introduction of IDEF0 modelling for process understanding and improvement.

Flow charting

In the systematic planning or examination of the detail of any sub-process, whether that be a clerical, manufacturing, or managerial activity, it is necessary to record the series of events and activities, stages and decisions in a form which can be easily understood and communicated to all. If the improvements are to be made, the facts relating to the existing method must be recorded first. The statements defining the process should lead to its understanding and will provide the basis of any critical examination necessary for the development of improvements. It is essential, therefore, that the descriptions of processes are accurate, clear and concise.

The usual method of recording facts is to write them down, but this is not suitable for recording the complicated sub-processes which exist in any organization. This is particularly so when an exact record is required of a long process, and its written description would cover several pages requiring careful study to elicit every detail. To overcome this difficulty certain methods of recording have been developed and the most powerful of these is flow charting. This method

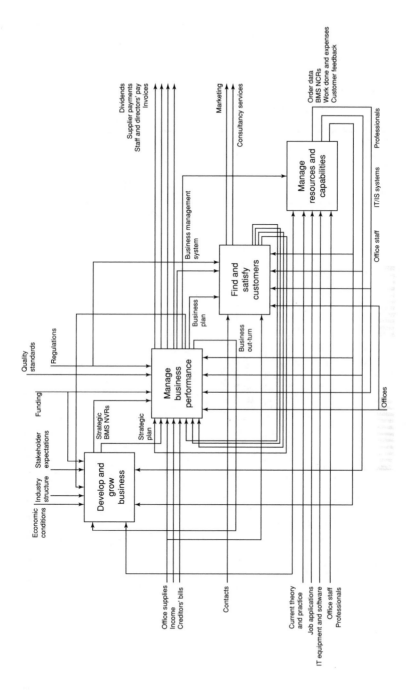

Figure 5.6 Process mapping at sub-process level

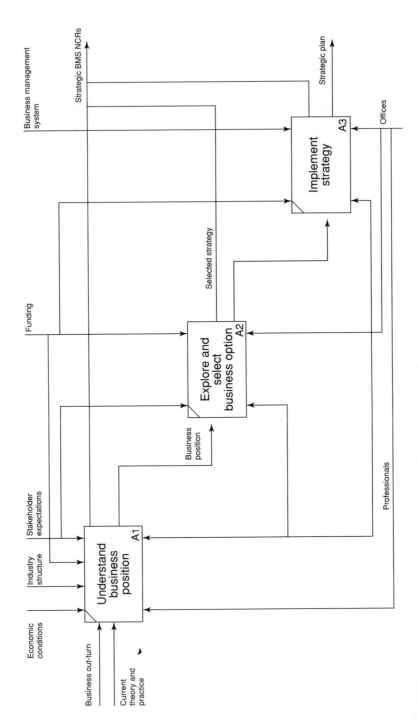

Figure 5.7 Process mapping at sub-sub-process level

of describing a process owes much to computer programming where the technique is used to arrange the sequence of steps required for the operation of the program. It has a much wider application, however, than computing.

Certain standard symbols are used on the chart and these are shown in Figure 5.8. The starting point of the process is indicated by a circle. Each processing step, indicated by a rectangle, contains a description of the relevant operation, and where the process ends is indicated by an oval. A point where the process branches because of a decision, is shown by a diamond. A parallelogram contains useful information but is not a processing step. The arrowed lines are used to connect symbols and to indicate direction of flow. For a complete description of the process all operation steps (rectangles) and decisions (diamonds) should be connected by pathways to the start circle and end oval. If the flow chart cannot be drawn in this way, the process is not fully understood.

It is a salutary experience for most people to sit down and try to draw the flow chart for a process in which they are involved every working day. It is often found that:

- the process flow is not fully understood,

- a single person is unable to complete the flow chart without help from others.

The very act of flow charting will improve knowledge of the process, and will begin to develop the teamwork necessary to find improvements. In many

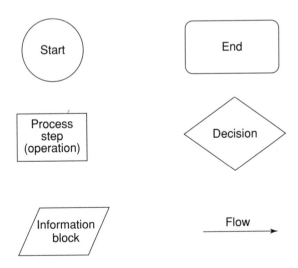

Figure 5.8   Flowcharting symbols

cases the convoluted flow and octopus like appearance of the chart will high-light unnecessary movement of people and materials and lead to common sense suggestions for waste elimination. An example of the use of flow chart-ing to improve a process is given in Figures 5.9 and 5.10, which show the 'before' and 'after' improvement of a travel arranging process.

It is surprisingly difficult to draw flow charts for even the simplest processes, particularly managerial ones, and following the first attempt it is useful to ask whether:

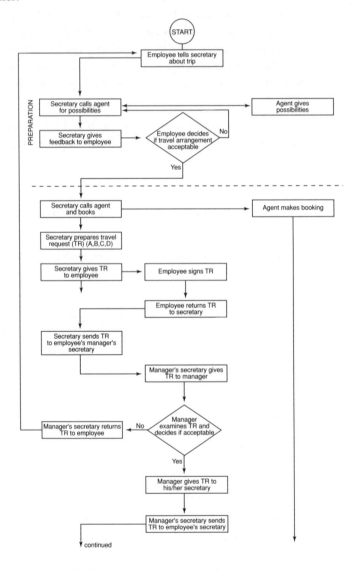

Figure 5.9 Original process for travel procedure

- the facts have been correctly recorded,

- any over-simplifying assumptions have been made,

- all the factors concerning the process have been recorded.

The author has seen too many process flow charts which are so incomplete as to be grossly inaccurate.

Flow charts may be of four basic types:

Person　　　　recording what people actually do;

Material　　　　recording how material (including paperwork) is handled or treated;

Equipment　　recording how equipment is used;

Information　　recording how information flows and to whom or where.

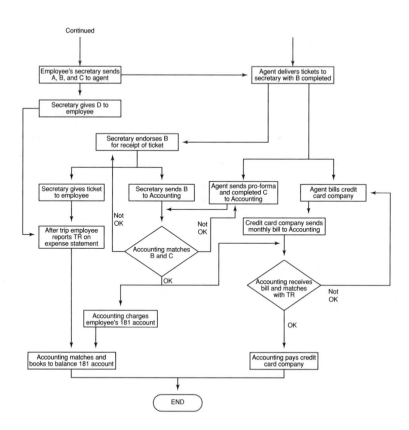

Figure 5.9 *(continued)*

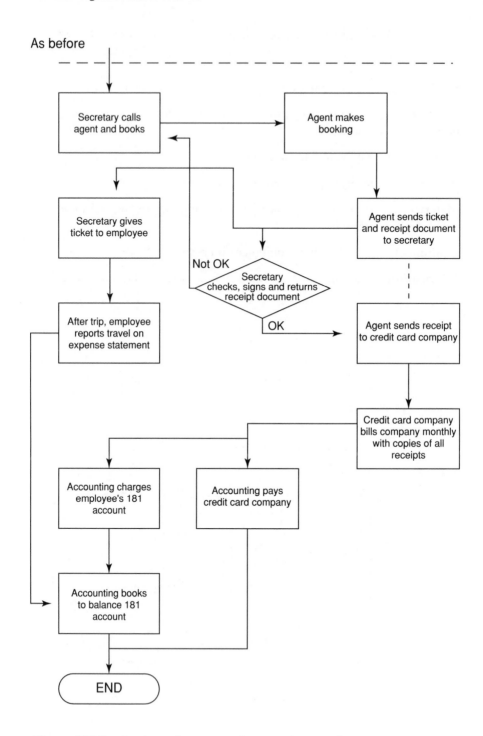

Figure 5.10 Improved process for travel procedure

Whichever type of flow chart is used, a critical examination is required, using a questioning technique which follows a well established sequence to examine:

the **purpose** for which ⎫
the **place** at which ⎪
the **sequence** in which ⎬ the activities are undertaken
the **people** by which ⎪
the **method** by which ⎭

with a view to

eliminating ⎫
combining ⎪
rearranging ⎬ those activities
or ⎪
simplifying ⎭

The questions which need to be answered in full are:

Purpose: What is actually done? (or what is actually achieved?)

Why is the activity necessary at all?

What else might be or should be done?

Eliminate unnecessary part of the job

Place: Where is it being done?

Why is it done at that particular place?

Where else might or should It be done?

Sequence: When is it done?

Why is it done at that particular time?

When might it or should it be done?

People: Who does it?

Why is it done by that particular person?

Who else might or should do it?

Combine wherever possible and/or **rearrange** operations for more effective results or reduction in waste

Method:	How is it done?	
	Why is it done in that particular way?	**Simplify** the operations
	How else might it or should it be done?	

Summarizing then, a flow chart is a picture of the steps used in performing a function. This function can be anything from a process step to accounting procedures, even preparing a meal. Lines connect the steps to show the flow of the various functions. Flow charts provide excellent documentation and are useful troubleshooting tools to determine how each step is related to the others. By reviewing the flow chart it is often possible to discover inconsistencies and determine potential sources of variation and problems. For this reason, flow charts are very useful in process improvement when examining an existing process to highlight the problem areas.

Process management and improvement – ten steps

The following are generic steps for mapping business processes:

1 Choose the process – based on identified areas for improvement, and determine its scope.

2 Choose the process sponsor – at the appropriate management level – and the process team – based on their involvement in the process.

3 Choose an outside facilitator – possibly a customer of or supplier to the process, or an external facilitator.

4 Decide on the title of the process (verb plus object) and the purpose of the process.

5 Determine the current level of performance of the process.

6 Identify the level of performance required for the process (also compare with best performance of a competitor or other outside company if possible/desirable).

7 IDEFO model high-level processes and sub-processes.

8 Flowchart the sub-processes, activities and tasks.

9 Determine and implement an action plan for process improvement or re-engineering.

10 Measure the results and continuously improve the process.

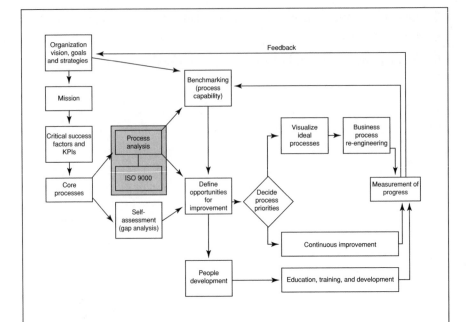

Process documentation and systems _____

Chapter 6

Process documentation and systems _____

Key points

An appropriate documented management system will help the objectives set out in the policy and strategy to be accomplished.

The International Standards Organization (ISO) 9000 and 14000 standards set out methods by which a system can be implemented to ensure that the specified requirements are met, and it is shown in this chapter how this should apply to and interact with all activities of the organization. The activities are generally processing, communicating, and controlling – these should be documented simply in the form of a manual.

The systems recommended follow the plan, do, check, act cycle of improvement, through documentation, implementation, audit and review.

The activities needed in the design and implementation of a good management system start and end with the customer, in two spheres – a customer sphere and a supplier sphere. It is stressed that senior management in all types of industry must take responsibility for the adoption and documentation of the appropriate management system in their organization if they are serious about achieving organizational excellence.

Why a documented system?

In earlier chapters we have seen how the keystone of organizational excellence is the concept of customer and supplier working together for their mutual advantage. For any particular organization, this becomes 'total' if the supplier/customer interfaces extend beyond the immediate customers, back inside the organization, and beyond the immediate suppliers. In order to achieve this, a company must organize itself in such a way that the human, administrative and technical factors affecting quality will be under control. This leads to the requirement for the development and implementation of a quality system which enables the objectives set out in the quality policy to be accomplished. Clearly, for maximum effectiveness and to meet individual customer requirements, the quality system in use must be appropriate to the type of activity and product or service being offered.

The aim of a good quality or management system is to provide the 'operator' of the process with consistency and satisfaction in terms of methods, materials, equipment, etc. Two feedback loops are also required, the 'voice' of the customer (marketing activities) and the 'voice' of the process (measurement activities).

This defines the way in which organizations must use the international standards on quality systems which are available. The 'wheel' has been invented but it must be built in a way which meets the specific organizational and product or service requirements. The International Standards Organization (ISO) Standard 9000 series sets out the methods by which a management system, incorporating all the activities associated with quality, can be implemented in an organization to ensure that all the specified performance requirements and needs of the customer are fully met.

Moreover the audit and review processes ensure that:

i) The people involved are operating according to the documented system (a system audit);

ii) The system still meets the requirements (a system review).

If the system audits and reviews discover that an even better product or less waste can be achieved by changing the method or one of the materials, then the management may wish to effect a change. To maintain consistency they must ensure that the appropriate changes are made to the documented system *and* that everyone concerned is issued with the revision and begins to operate accordingly.

A fully documented quality or management system will ensure that two major requirements are met:

i) The customer's requirements – for confidence in the ability of the organization to deliver consistently the desired product or service.

ii) The organization's requirements – both internally and externally, and at an optimum cost, with efficient utilization of the resource available – material, human, technological and administrative.

These requirements can be truly met only if objective evidence is provided in the form of information and data, which supports the system activities, from the ultimate supplier through to the ultimate customer.

A **quality management system** may be defined then as an assembly of components, such as the organizational structure, responsibilities, procedures, and resources for implementing organizational excellence. These components interact together and are affected by being in the system, so the isolation and study of each one in detail will not necessarily lead to an understanding of the system as a whole. Often the interactions between the components – such as materials and processes, procedures and responsibilities – are just as important as the components themselves, and problems can arise from these interactions as much as from the components. Clearly if one of the components is removed from the system, the whole thing will change.

Quality management system design

The quality management system should apply to and interact with all activities of the organization. This begins with the identification of the requirements and ends with their satisfaction, at every transaction interface. The activities involved may be classified in several ways – generally as processing, communicating and controlling, but more usefully and specifically as:

1   marketing

2   market research

3   design

4   specification

5   development

6   procurement

7   process planning

8   process development and assessment

9   process operation and control

10   product or service testing or checking

11   packaging (if required)

12   storage (if required)

13   sales

14   distribution or installation/operation

15   technical service

16   maintenance/customer service

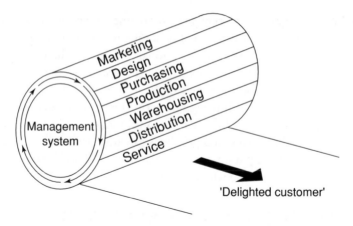

'Delighted customer'

Figure 6.1 Management system power unit to the 'delighted customer'

These may be regarded as slats on a rotating drum rolling towards a satisfied customer, who becomes 'delighted' by the consistency of the product or service (Figure 6.1). The driving force of the drum is in the centralized management system and the drum will not operate until the system is in place and working. The first step in getting the drum rolling is to prepare the necessary documentation. This means, in very basic terms, that procedures should be written down, preferably in such a way that the system conforms to one of the national or international standards. This is probably best done in the form of a quality or management manual.

It is interesting to bring together the concepts of Deming's cycle of continuous improvement, plan–do–check–act, and quality management systems. A simplification of what a good quality system is trying to do is given in Figure 6.2, which follows the improvement cycle.

In most organizations, established methods of working already exist and all that is required is the **writing down of what is currently done**. In some instances, companies may not have procedures to satisfy the requirements of a good standard and they may have to begin to devise them. Alternatively, it may be found that two people, supposedly performing the same task, are working in different ways, and there is a need to standardize the procedure. Some organizations use the effective slogan, 'if it isn't written down, it doesn't exist'. This can be a useful discipline provided it doesn't lead to bureaucracy.

Justify that the **system** as it is designed **meets the requirements of a good international standard,** such as ISO 9000. There are other excellent standards

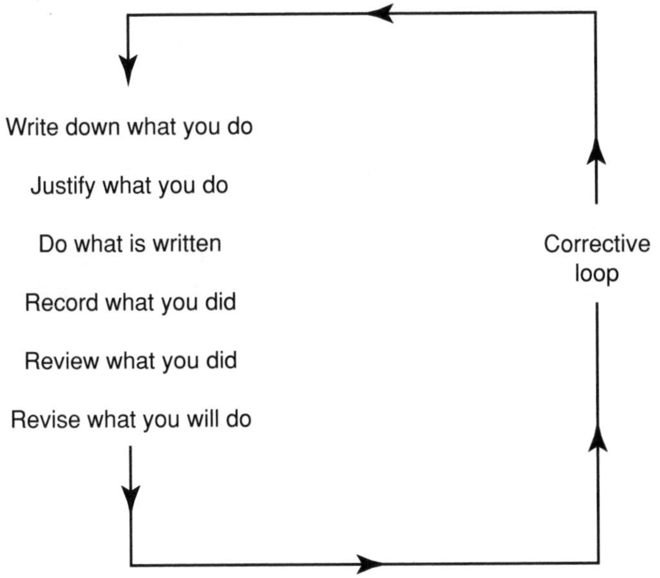

Figure 6.2 The quality system and never-ending improvement

which are used, e.g. QS 9000, and these provide similar 'check lists', of things to consider in the establishment of the quality system.

Make sure that the system is a working one and the documents are well fingered in use. One person alone cannot document a quality system, the task must involve everyone who has responsibility for any part of it. The quality manual must be a **practical working document** – that way it ensures that consistency of operation is maintained and it may be used as a training aid.

In the operation of any process, a useful guide is:

- No process without data collection.

- No data collection without analysis.

- No analysis without decisions.

- No decisions without actions, which can include doing nothing.

This excellent discipline is built into any good quality management system, primarily through the audit and review systems. The requirement to **audit or 'check'** that the system is functioning according to plan, and to **review** possible system improvements, utilizing audit results, should ensure that the **improvement cycle** is engaged through the **corrective action** procedures.

The overriding requirement is that the systems must reflect the established practices of the organization, improved where necessary to bring them into line with current and future requirements.

Environmental management systems

Organizations of all kinds are increasingly concerned to achieve and demonstrate sound environmental performance. Many have undertaken environmental audits and review to assess this. To be effective these need to be conducted within a structured management system, which in turn is integrated with the overall management activities dealing with all aspects of desired environmental performance.

Such a system should establish procedures for setting environmental policy and objectives, and achieving compliance to them. It should be designed to place emphasis on the prevention of adverse environmental effects, rather than on detection after occurrence. It should also identify and assess the environmental effects arising from the organization's existing or proposed activities, products, or services and from incidents, accidents, and potential emergency situations. The system must identify the relevant regulatory requirements, the priorities, and pertinent environmental objectives and targets. It needs also to facilitate planning, control, monitoring, auditing and review activities to ensure that the policy is complied with, that it remains relevant, and is capable of evolution to suit changing circumstances.

The international standard ISO 14000 contains a specification for environmental management systems for ensuring and demonstrating compliance with stated policies and objectives. The standard is designed to enable any organization to establish an effective management system, as a foundation for both sound environmental performance and participation in environmental auditing schemes.

ISO 14000 shares common management system principles with the ISO 9000 series and organizations may elect to use an existing management system, developed in conformity with the ISO 9000 series, as a basis for environmental management. The ISO 14000 standard defines environmental policy, objectives, targets, effect, management, systems, manuals, evaluation, audits and reviews. It mirrors the ISO 9000 series requirements in many of its own requirements, and it includes a guide to these in an informative annex.

Management systems are needed in all areas of activity, whether large or small businesses, manufacturing, service or public sector. The advantages of systems in manufacturing are obvious, but they are just as applicable in areas such as marketing, sales, personnel, finance, research and development, as well as in the service industries and public sectors.

No matter where it is implemented a good management system will improve process control, reduce wastage, lower costs, increase market share (or funding), facilitate training, involve staff, and raise morale.

Internal and external management system audits and reviews

A good management system will not function without adequate audits and reviews. The system reviews, which need to be carried out periodically and systematically, are conducted to ensure that the system achieves the required effect, whilst audits are carried out to make sure that actual methods are adhering to the documented procedures. The reviews should use the findings of the audits, for failure to operate according to the plan often signifies difficulties in doing so. A re-examination of the procedures actually being used may lead to system improvements unobtainable by other means.

A schedule for carrying out the **audits** should be drawn up, different activities perhaps requiring different frequencies. All procedures and systems should be audited at least once during a specified cycle, but not necessarily all at the same audit. For example, every three months a selected random sample of work instructions and test methods could be audited, with the selection designed so that each procedure is audited at least once per year. There must be, however, facility to adjust this on the basis of the audit results.

A quality system **review** should be instituted, perhaps every six months, with the aims of:

- ensuring that the system is achieving the desired results,
- revealing defects or irregularities in the system,
- indicating any necessary improvements and/or corrective actions to eliminate waste or loss,
- checking on all levels of management,
- uncovering potential danger areas,
- verifying that improvements or corrective action procedures are effective.

Clearly, the procedures for carrying out the audits and reviews and the results from them should be documented, and themselves be subject to review.

The assessment of a quality system against a particular standard or set of requirements by internal audit and review is known as a **first party** assessment or approval scheme. If an **external** customer makes the assessment of a supplier against either its own or a national or international standard, a **second party** scheme is in operation. The assessment by an independent organization, not involved in any contract between customer and supplier, but acceptable to them both, is known as an **independent third party** assessment scheme. The latter usually involves some form of certification or registration by the assessment body.

One advantage of the third party schemes is that they obviate the need for customers to make their own detailed checks, saving both suppliers and customers time and money, and avoiding issues of commercial confidentiality. Just one knowledgeable organization has to be satisfied, rather than a multitude with varying levels of competence. This often qualifies suppliers for quality assurance based contracts without further checking.

Each certification body usually has its own recognized mark, which may be used by registered organizations of assessed capability in their literature, letter headings, and marketing activities. There are also publications containing lists of organizations whose quality systems and/or products and services have been assessed. To be of value, the certification body must itself be recognized and, usually, assessed and registered with a national or international accreditation scheme, such as the UK Council for Certification Bodies (UCCB).

Many organizations have found that the effort of designing and implementing a written management system, good enough to stand up to external independent third party assessment, has been extremely rewarding in:

• involving staff and improving morale,

• better process control,

• reduced wastage,

• reduced customer service costs.

This is also true of those organizations that have obtained third party registration and supply companies which still insist on their own second party assessment. The reason for this is that most of the standards on management systems, whether national, international, or company specific, are now very similar indeed. A system which meets the requirements of the ISO 9000 series, for example, should meet the requirements of all other standards, with only the slight modifications and small emphases here and there required for specific customers. It is the author's experience, and that of his immediate colleagues, that an assessment carried out by one of the independent certified assessment bodies is at least as rigorous and delving as any carried out by a second party representative.

Internal system audits and reviews must be positive and conducted as part of the preventive strategy and not as a matter of expediency resulting from quality problems. They should not be carried out only prior to external audits, nor should they be left to the external auditor – whether second or third party. An external auditor, discovering discrepancies between actual and documented systems, will be inclined to ask why the internal review methods did not discover and correct them. This type of behaviour in financial control and auditing is commonplace.

Managements should become fully committed to operating an effective management system which involves all personnel within the organization, not just the staff in the 'quality department'. The system must be planned to be effective and achieve its objectives in an uncomplicated way. Having established and documented the procedures it is necessary to ensure that they are working and that everyone is operating in accordance with them. The system once established is not static; it should be flexible to enable the constant seeking of improvements or streamlining.

Quality auditing standard

The growing use of standards internationally emphasizes the importance of auditing as a management tool for this purpose. There are available several guides to quality systems auditing and the guidance provided in these can be applied equally to any one of the three specific and yet different auditing activities:

i) **First party or internal audits**, carried out by an organization on its own systems, either by staff who are independent of the systems being audited, or by an outside agency.

ii) **Second party audits,** carried out by one organization (a purchaser or its outside agent) on another with which it either has contracts to purchase goods or services or intends to do so.

iii) **Third party audits,** carried out by independent agencies, to provide assurance to existing and prospective customers for the product or service.

Audit objectives and responsibilities, including the roles of auditors and their independence, and those of the 'client' or auditee should be understood. The generic steps involved then are as follows:

• **initiation**, including its scope and frequency;

- **preparation**, including review of documentation, the programme, and working documents;

- **execution**, including the opening meeting, examination and evaluation, collecting evidence, observations, and closing the meeting with the auditee;

- **report**, including its preparation, content and distribution;

- **completion**, including report submission and retention.

Attention should be given at the end of the audit to corrective action and follow-up and the improvement process should be continued by the auditee after the publication of the audit report. This may include a call by the client for a verification audit of the implementation of any corrective actions specified.

The rings of confidence

The activities which must be addressed in the design and implementation of a good management system may be considered to be attached to a 'ring of confidence', which starts and ends with the customer (Figure 6.3).

It is possible to group these into two spheres of activities:

- those involving direct interaction with the customer,

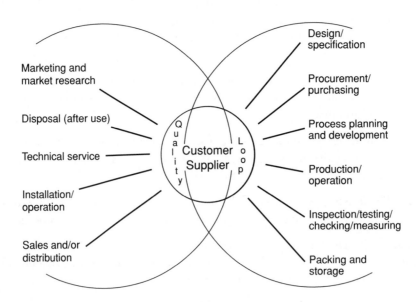

Figure 6.3 The rings of confidence

- those concerning primarily the internal activities of the supplier.

The overlap necessary between customer and supplier is clearly illustrated by this model. Equally obvious is that separation will lead to dysfunction and dissatisfaction.

It cannot be stated too often that the customer–supplier interactions, which generate satisfaction of needs, are just as necessary internally. The principles of management system design, documentation and implementation set out in this chapter must apply to every single person, every department, every process transaction, and every type of organization. The vocabulary in the engineering factory system may be different from that used in the hotel; the hospital system will be set out differently to that of the drug manufacturer, but the underlying concepts will be the same.

It is not acceptable for the managers in industries, or parts of organizations, less often associated with standards on management systems to find 'technological' reasons for avoiding the requirement to manage. The author and his colleagues have heard the excuse that 'our industry (or organization) differs from any other industry (or organization)', in almost every industry or organization with which they have been involved. Clearly, there are technological differences between all industries and nearly all organizations, but in terms of managing processes there are hardly any at all.

Senior managers in every type and size of organization must take the responsibility for the adoption of the appropriate documented management system. If this requires translation from 'engineering language', so be it – get someone from inside or outside the organization to do it. Do not wait for the message to be translated into different forms – inefficiencies, waste, high costs, crippling competition, loss of market.

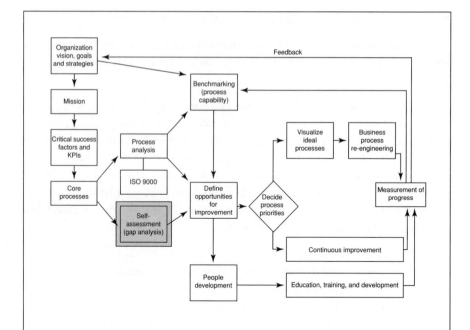

Self-assessment – gap analysis _____

Chapter 7

Self-assessment –
gap analysis _____

Key points

This chapter discusses the most famous and widely used frameworks for self-assessment in the West; the Malcolm Baldrige National Quality Award (BNQA) in the USA and the European/UK Business Excellence Award.

The MBNQ criteria are built on a set of core values and concepts which are embodied on a framework of seven first level categories: leadership, strategic planning, customer and market focus, information and analyses, human resource development, process management, and business results. These are comparable with the categories of the Japanese Deming Prize, and the nine components of the European Excellence Award: leadership driving policy and strategy through management of people, processes and resources and partnerships, delivering superior people results, customer results and societal results, which in turn lead to better key performance results. The various award criteria provide rational bases against which to measure progress towards excellence in organizations and the so-called 'self-assessment' methods currently in use are explained.

It is shown how the 'motors' driving an organization towards its mission must be linked to its five stakeholders: customers, employees, suppliers (determinants) and shareholders, community (resultants).

Frameworks for self-assessment

Organizations everywhere are under constant pressure to improve their business performance, measure themselves against world class standards and focus their efforts on the customer. To help in this process, many are turning to total quality models such as the European Foundation for Quality Management's (EFQM) 'Excellence Model' – promoted in the UK by the British Quality Foundation.

'Total quality' is the goal of many organizations but it has been difficult until relatively recently to find a universally accepted definition of what this actually means. For some people TQ means statistical process control (SPC) or quality systems, for others teamwork and involvement of the workforce.

Clearly there are many different views on what constitutes the 'excellent' organization and, even with an understanding of a framework, there exists the difficulty of calibrating the performance or progress of any organization towards it.

'Business excellence' recognizes that customer results, business objectives, safety, and environmental considerations are mutually dependent, and that it is applicable in any organization. Clearly the application of the ideas involves investment, primarily in people and time, time to implement new concepts, time to train, time for people to recognize the benefits and move forward into new or different organizational cultures. But how will organizations know when they are getting close to excellence or whether they are even on the right road, how will they measure their progress and performance?

There have been many recent developments and there will continue to be many more, in the search for a standard or framework, against which organizations may be assessed or measure themselves, and carry out the so-called 'gap analysis'. To many the ability to judge progress against an accepted set of criteria would be most valuable and informative.

The Baldrige National Quality Award criteria (USA)

Most TQ approaches strongly emphasize measurement, especially in the quality assurance and control areas. Some insist on the use of cost of quality. The recognition that total quality management is a broad culture change vehicle with internal and external focus embracing behavioural and service issues, as well as quality assurance and process control, prompted the United States to develop one of the most famous and now widely used frameworks, the (Malcolm) Baldrige National Quality Award (BNQA). The award itself, which is composed of two solid crystal forms 14 inches high, is presented annually to recognize companies in the USA that have 'excelled in quality management and quality achievement'. But it is not the award itself, or even the fact that it is presented each year by the President of the USA, which has attracted the attention of most organizations. It is the excellent framework, which is one of the closest things we have to an international standard for total quality.

The value of a structured discipline using a points system has been well established in quality and safety assurance systems (for example, ISO 9000 and vendor auditing). The extension of this approach to a total quality auditing process has been long established in the Japanese Deming Prize which is perhaps the most demanding and intrusive auditing process, and there are other excellent models and standards used throughout the world.

In 1987 the BNQA was introduced for United States based organizations. Many companies have realized the necessity to assess themselves against the Baldrige criteria, if not to enter for the Baldrige Award then certainly as an excellent basis for self-audit and review, to highlight areas for priority attention and provide internal and external benchmarking.

The BNQA aims to promote:

- understanding of the requirements for performance excellence and competitiveness improvement, and
- sharing of information of successful performance strategies and the benefits to be derived from using these strategies.

The award criteria are built upon a set of core values and concepts:

- customer-driven quality
- leadership
- continuous improvement and learning
- valuing employees
- fast response
- design quality and prevention
- long-range view of the future
- management by fact
- partnership development
- company responsibility and citizenship
- results focus.

These are embodied in a framework of seven first level categories which are used to assess organizations:

- Leadership;
- Strategic planning;
- Customer and market focus;
- Information and analysis;
- Human resource focus;

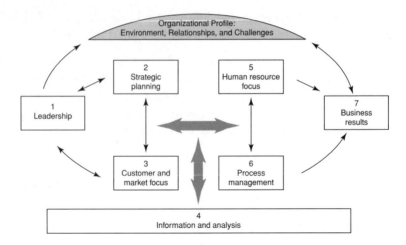

Figure 7.1   Baldrige criteria for performance excellence framework: a systems perspective.
Source: Malcolm Baldrige National Quality Award, 'Criteria for Performance Excellence', 2001, US National Institute of Standards and Technology, Gaithersburg

- Process management;
- Business results.

Figure 7.1 shows how the framework's system connects and integrates the categories. This has three basic elements: strategy and action plans (customer and market focused), system, and information and analyses. The main driver is the senior executive leadership which creates the values, goals and systems, and guides the sustained pursuit of excellence and performance objectives. The system includes a set of well defined and designed processes for meeting the organization's direction and performance requirements. Measures of progress provide a results oriented basis for channelling actions to deliver ever improving customer values and organization performance. The overall goal is the delivery of customer satisfaction and market success leading, in turn, to excellent business results. The seven criteria categories are further subdivided into items and areas to address. These are described in some detail in the 'Criteria for Performance Excellence' available from the US National Institute of Standards and Technology in Gaithersburg (www.quality.nist.gov).

The European Quality Award

In Europe it has also been recognized that the technique of self-assessment is very useful for any organization wishing to monitor and improve its performance. In

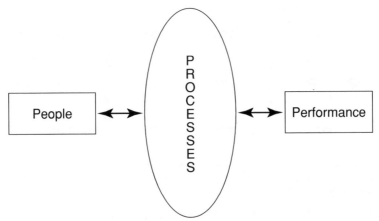

Achieve better performance through involvement of all
employees (people) in continuous improvement
of their processes

Figure 7.2   The simple model for improved performance

1992 the European Foundation for Quality Management (EFQM) launched a
European Quality Award framework which is now widely used for systematic
review and measurement of operations. The EFQM 'Excellence Model' recognized that processes are the means by which a company or organization har-
nesses and releases the talents of its people to produce results performance.
Moreover, improvement in the performance can be achieved only by improving
the processes by involving the people. This simple model is shown in Figure 7.2.

Figure 7.3 displays graphically the principle of the full EFQM Excellence
Model.* Essentially customer results, people results, and favourable society
results are achieved through leadership driving policy and strategy, people,
resources and partnerships, and processes, which lead ultimately to excellence
in key performance results – the enablers deliver the results. The EFQM have
provided a weighting for each criteria, shown in Figure 7.3, which is now
widely used in scoring self-assessments and making awards. The weightings are
not rigid and may be modified to suit specific organizational needs.

The EFQM have thus built a model of criteria and a review framework
against which an organization may face and measure itself, to examine any
'gaps'. Such a process is known as self-assessment and organization such as

---

* An excellent resource is 'The Model in Practice – using the EFQM Excellence Model to deliver continuous improvement' published by BQF, 2001.

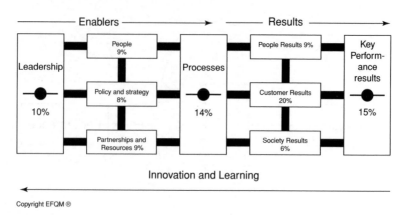

Copyright EFQM ®

Figure 7.3 The model for business excellence

EFQM and in the UK, the British Quality Foundation (BQF) publish guidelines for self-assessment, including specific ones directed at public sector organizations.

Many managers feel the need for a rational basis on which to measure progress in their organization, especially in those companies a few years 'into TQM' which would like the answers to questions such as: 'Where are we now?' 'Where do we need/want to be?' and 'What have we got to do to get there?' These questions need to be answered from internal employees' views, the customers' views, and the views of suppliers.

Gap analyses using self-assessment to the EFQM Excellence Model promotes business excellence through a regular and systematic review of business processes and results. It highlights strengths and improvement opportunities, and drives continuous improvement.

Enablers

In the EFQM Excellence Model, the enabler criteria of: leadership, policy and strategy, people, resources and partnerships, and processes focus on what is needed to be done to achieve results. The structure of the enabler criteria is shown in Figure 7.4. Enablers are assessed on the basis of the combination of two factors (see Figure 7.5, Chart 1, The enablers):

1 The degree of excellence of the approach

2 The degree of deployment of the approach

The detailed criterion parts are as follows:

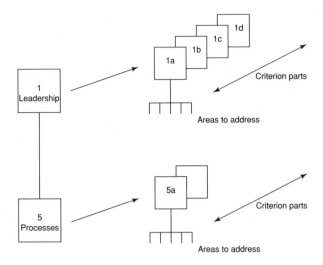

Figure 7.4   Structure of the criteria: enablers

1   **Leadership**

How leaders develop and faciliate the achievement of the mission and vision, develop values rerquired for long term success and implement these via appropriate actions and behaviours, and are personally involved in ensuring that the organization's 'management' system is developed and implemented.

Self-assessment should demonstrate how leaders:

1a)   develop the mission, vision and values and are role models of a culture of excellence
1b)   are personally involved in ensuring the organization's management system is developed, implemented and continuously improved
1c)   are involved with customers, partners and representatives of society
1d)   motivate, support and recognize the organization's people

2   **Policy and strategy**

How the organization implements its mission and vision via a clear stake-holder focused strategy, supported by relevant policies, plans, objectives, targets and processes

Self-assessment should demonstrate how policy and strategy are:

2a)   based on the present and future needs and expectations of stake-holders
2b)   based on information from performance measurement, research, learning and creativity-related activities

Approach	Score	Deployment, Assessment and Review
Anecdotal or no evidence.	0%	Little effective usage.
Some evidence of soundly based approaches and prevention based processes/systems.	25%	Implemented in about one-quarter of the relevant areas and activities
Some evidence of integration into normal operations.		Some evidence of assessment and review
Evidence of soundly based systematic approaches and prevention based processes/systems.	50%	Implemented in about half the relevant areas and activities.
Evidence of integration into normal operations and planning well established.		Evidence of assessment and review
Clear evidence of soundly based systematic approaches and prevention based processes/systems.	75%	Applied to about three-quarters of the relevant areas and activities.
Clear evidence of integration of approach into normal operations and planning.		Clear evidence of refinement and improved business effectiveness through review cycles.
Comprehensive evidence of soundly based systematic approaches and prevention based processes/systems.	100%	Implemented in all relevant areas and activities.
Approach has become totally integrated into normal working patterns. Could be used as a role model for other organizations.		Comprehensive evidence of refinement and improved business effectiveness through review cycles.

For **Approach**, **Deployment**, **Assessment** and **Review** the assessor may choose one of the five levels 0%, 25%, 75%, or 100% as presented in the chart, or interpolate between these values.

Figure 7.5 Scoring within the self-assessment process: Chart 1, The enablers

2c) developed, reviewed and updated

2d) deployed through a framework of key processes

2e) are communicated and implemented

3 People

How the organization manages, develops and releases the knowledge and full potential of its people at an individual, team-based and organization-wide level, and plans these activities in order to support its policy and strategy and effective operation of its processes.

Self-assessment should demonstrate how people:

3a) resources are planned, managed and improved

3b) knowledge and competences are identified, developed and sustained

3c) are involved and empowered

3d) and the organization have a dialogue

3e) are rewarded, recognized and cared for.

4 Partnerships and resources

How the organization plans and manages its external partnerships and internal resources in order to support its policy and strategy and the effective operation of its processes.

Self-assessment should demonstrate how:

4a) external partnerships are managed

4b) finances are managed

4c) buildings, equipment and materials are managed

4d) technology is managed

4e) information and knowledge are managed

5 Processes

How the organization designs, manages and improves its processes in order to support its policy and strategy and fully satisfy, and generate increasing value, for its customers and other stakeholders.

Self-assessment should demonstrate how:

5a) processes are designed and systematically managed

5b) processes are improved, as needed, using innovation in order to fully satisfy and generate increasing value for customers

5c) products and services are designed and developed based on customer needs and expectations

5d) products and services are produced, delivered and serviced

5e) customer relationships are managed and enhanced

Assessing the enablers criteria

These criteria are concerned with how an organization or business unit achieves its results. Self-assessment asks the following questions in relation to each criterion part:

- What is currently done in this area?

- How is it done? Is the approach systematic and prevention based?

- How is the approach reviewed and what improvements are undertaken following review?

- How widely used are these practices?

Results

The EFQM Excellence Model's result criteria of: customer results, people results, society results, and key performance results focus on what the organization has achieved and is achieving. These can be expressed as discrete results, but ideally as trends over a period of years. The structure of the results criteria is shown in Figure 7.6.

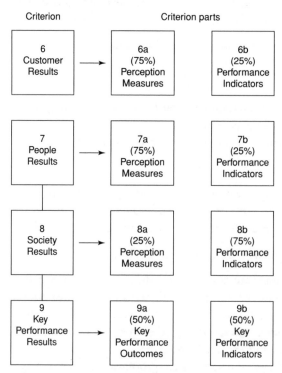

Figure 7.6   Structure of the criteria: results

'Performance excellence' is assessed relative to the organization's business environment and circumstances, based on information which sets out:

- the organization's actual performance
- the organization's own targets

and wherever possible:

- the performance of competitors or similar organizations
- the performance of 'best in class' organizations

Results are assessed on the basis of the combination of two factors (see Figure 7.7, Chart 2, The results):

1   The degree of excellence of the results
2   The scope of the results

The detailed assessment of results is as follows:

6   **Customer results**

What the organization is achieving in relation to its external customers.

Self-assessment should demonstrate the organization's success in satisfying the needs and expectations of its external customers.

Areas to consider:
6a) Results achieved for the measurement of customer perception of the organization's products, services and customer relationships
6b) Internal performance indicators relating to the organization's customers.

7   **People results**

What the organization is achieving in relation to its people

Self-assessment should demonstrate the organization's success in satisfying the needs and expectations of its people.

Areas to consider:
7a) Results of people's perception of the organization
7b) Internal performance indicators relating to people

8   **Society results**

What the organization is achieving in relation to local, national and international society as appropriate

Results	Score	Scope
No results or anecdotal information.	0%	Results address few relevant areas and activities.
Some results show positive trends and/or satisfactory performance. Some favourable comparisons with own targets/external organizations. Some results are caused by approach	25%	Results address some relevant areas and activities.
Many results show strongly positive trends and/or sustained good performance over at least 3 years. Favourable comparisons with own targets in many areas. Some favourable comparison with external organizations. Many results are caused by approach.	50%	Results address many relevant areas and activities.
Most results show strong positive trends and/or sustained excellent performance over at least 3 years. Favourable comparisons with own targets in most areas. Favourable comparisons with external organizations in many areas. Most results are caused by approach.	75%	Results address most relevant areas and activities.
Strongly positive trends and/or sustained excellent performance in all areas over at least 5 years. Excellent comparisons with own targets and external organizations in most areas. All results are clearly caused by approach. Positive indication that leading position will be maintained.	100%	Results address all relevant areas and facets of the organization.

For both **Results** and **Scope**, the assessor may choose one of the five levels 0%, 25%, 50%, 75%, or 100% as presented in the chart, or interpolate between these values.

Figure 7.7 Scoring within the self-assessment process: Chart 2, The results

Self-assessment should demonstrate the organization's success in satisfying the needs and expectations of the community at large.

Areas to consider:

8a) Society's perception of the organization
8b) Internal performance indicators relating to the organization and society

9 **Key performance results**

What the organization is achieving in relation to its planned performance.

Areas to consider:

9a) Key performance outcomes, including financial and non-financial
9b) Key indicators of the organization's performance which might predict likely key performance outcomes

Assessing the results criteria

These criteria are concerned with what an organization has achieved and is achieving. Self-assessment addresses the following issues:

- The measures used to indicate performance
- The extent to which the measures cover the range of the organization's activities
- The relative importance of the measures presented
- The organization's actual performance
- The organization's performance against targets

and wherever possible:

- Comparisons of performance with similar organizations
- Comparisons of performance with 'best in class' organizations

Methodologies for self-assessment

The EFQM provide a flow diagram of the general steps involved in undertaking self-assessment. A simplified version of this is shown in Figure 7.8.

There are a number of approaches to carrying out self-assessment including:

- discussion group/workshop methods
- surveys, questionnaires and interviews (peer involvement)

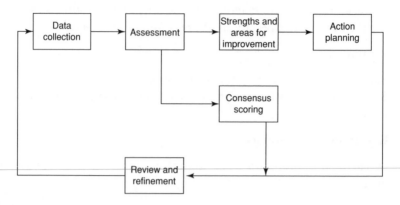

Figure 7.8 The key steps in self-assessment

- pro formas
- organizational self-analysis matrices (e.g. see Figure 7.9.)
- an award simulation
- activity or process audits
- hybrid approaches

Whichever method is used, the emphasis should be on understanding the organization's strengths and areas for improvement, rather than the score. The scoring charts provide a consistent basis for establishing a quantitative measure of performance against the model and gaining consensus promotes discussion and development of the issues facing the organization. It should also gain the involvement, interest and commitment of the senior management, but the scores should not become an end in themselves. Tito Conti, often called 'the father of self-assessment', following the contribution he made to its establishment and development through the EFQM, when he was head of Fiat, has expressed concern that organizations can become obsessed with self-assessment scores rather than focusing on the improvement opportunities identified.

A few last words on self-assessment

There is great overlap between the criteria used by the various awards and it may be necessary for an organization to rationalize them. The main components, however, must be the organization's processes, management systems, people management and results, customer results and performance results. Self-assessment provides an organization with vital information in monitoring its progress towards its goals and 'business excellence'. The external assessments used in the processes of making awards must be based on these self-assessments that are performed as prerequisites for improvement.

Whatever are the main 'motors' for driving an organization towards its vision or mission, they must be linked to the five stakeholders embraced by the values of any organization, namely: customers, employees, suppliers, shareholders and community.

In any normal business or organization, measurements are continuously being made, often in retrospect, by the leaders of the organization to reflect the value put on the organization by its five stakeholders. Too often, these continuous readings are made by internal biased agents with short term priorities, not always in the best long term interests of the organization or its customers, Third party agents, however, can carry out or facilitate periodic audits and reviews from the perspective of one or more of the key stakeholders, with particular emphasis on forward priorities and needs. These reviews will allow realignment of the principle driving motors to focus on the critical success factors and continuous improvement, to maintain a balanced and powerful general thrust which moves the whole organization towards its mission.

The relative importance of the five stakeholders may vary in time but all are important. The first three, customers, employees, and suppliers, which comprise the core value chain, are the determinant elements. The application of total quality principles in these areas will provide satisfaction as a resultant to the shareholders and the community. Thus, added value will benefit the community and the environment. The ideal is a long way off in most organizations, however, and active attention to the needs of the shareholders and/or community remain a priority for one major reason – they are the 'customers' of most organizational activities and are vital stakeholders.

Any instrument which is developed for self-assessment may be used at several stages in an organization's history:

- before starting an improvement programme to identify 'strengths' and 'areas for improvement', and focus attention.

- as part of a programme launch, especially using a 'survey' instrument

- every one or two years after the launch to steer and benchmark.

The systematic measurement and review of operations are some of the most important management activities of any organization. Self-assessment leads to clearly discerned strengths and areas for improvement by focusing on the relationship between the people, processes, and performance. Within any successful organization it will be a regular activity.

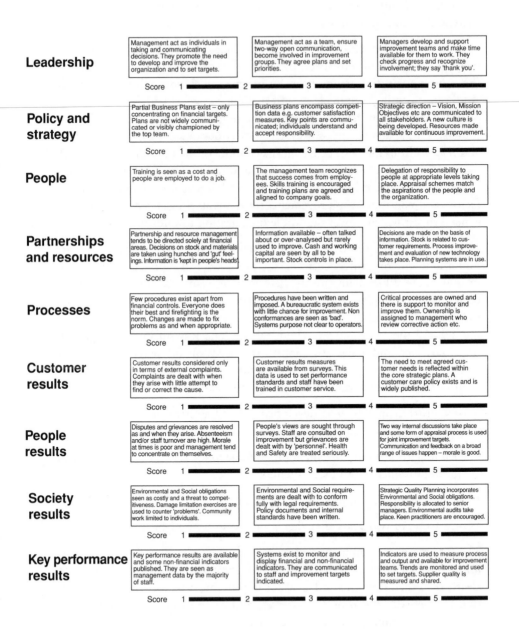

Figure 7.9 Organizational self-analysis matrix
Source: UK North West Quality Award Model

Starting with Leadership, read all the statements across the page and choose and circle the number you feel best reflects the situation within your organization. Multiply the number you have chosen by the factor shown and enter your score in the box on the right. Repeat the exercise for the other eight criteria and total your scores to produce a Grand Total.

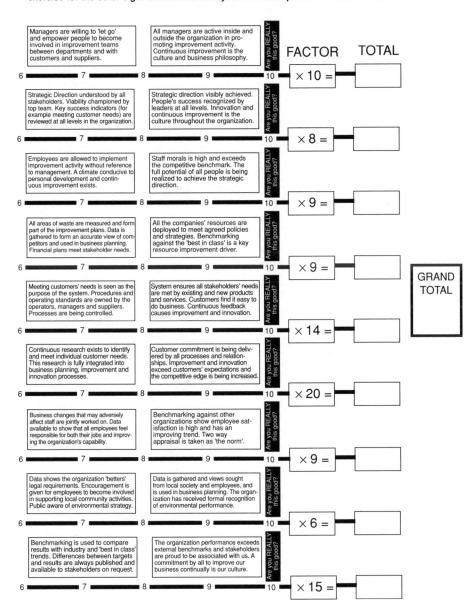

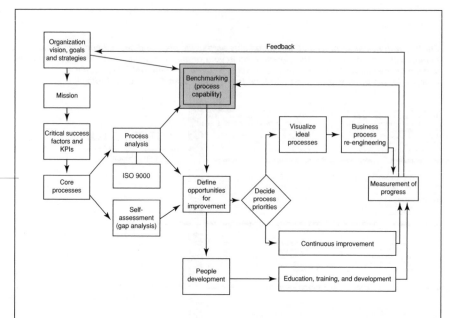

Benchmarking _____

Chapter 8

Benchmarking _____

Key points

This chapter will show how benchmarking measures an organization's products, services and processes to establish targets, priorities and improvements, leading in turn to competitive advantage and/or cost reductions.

The four basic types of benchmarking are: internal, competitive, functional and generic, although the evolution of benchmarking in an organization is likely to progress through focus on continuous improvement.

Benchmarking is presented as an efficient way to promote effective change, by learning from the successful experiences of others and putting that learning to good effect.

A strategy for the execution of benchmarking, with five main stages of plan, analyse, develop, improve and review, is given.

Benchmarking – what is it and what are the benefits?

Product, service and process improvements can take place only in relation to established standards, with the improvements then being incorporated into the new standards. Benchmarking, one of the most transferable aspects of Rank Xerox's approach to total quality management, thought to have originated in Japan, measures an organization's operations, products and services against those of its competitors and others in a ruthless fashion. It is a means by which targets, priorities and operations can be established that will lead to competitive advantage.

The concept is based on the ancient Japanese quotation: 'If you know your enemy and know yourself, you need not fear the result of a hundred battles.' (Sun Tzu, *The Art of War*, 500BC). In Japanese, the word *dantoso* means striving for the best of the best.

The word 'benchmark' is a reference or measurement standard used for comparison, and benchmarking is the continuous process of identifying, understanding and adapting best practice and processes that will lead to superior performance.

Benchmarking is *not* primarily an exercise in cost reduction or in gathering data which can be used as a political weapon. Nor is it a panacea to cure all ills, a public relations exercise, an operation to be taken lightly. Organizations use benchmarking generally as a stimulus to stay or get ahead and as a means of managing the changes needed to maximize the benefits. Of course, your customers 'benchmark' every time they telephone your organization.

There may be many reasons for carrying out benchmarking. Some of them are set against various objectives in Table 8.1. The links between benchmarking and excellence are clear – establishing objectives based on best practice should directly contribute to better meeting of the internal and external customer requirements.

The benefits of benchmarking can be numerous but include:

- creating a better understanding of the current position

- heightening sensitivity to changing customer needs

- encouraging innovation

- developing realistic stretch goals

- establishing realistic action plans

Data from the American Productivity and Quality Center's International Benchmarking Clearinghouse suggests that an average benchmarking study takes six months to complete, occupies more than a quarter of the team members' time, and costs around £50 000. The same source identified that the average return was **five times** the cost of the study, in terms of reduced costs, increased sales, greater customer retention and enhanced market share.

Benchmarking in practice

There are four basic types of benchmarking:

Internal   a comparison of internal operations and processes,

Competitive   specific competitor to competitor comparisons for a product or function of interest,

Functional   comparisons to similar functions within the same broad industry or to industry leaders,

Generic   comparisons of business processes or functions that are very similar, regardless of the industry.

The evolution of benchmarking in an organization is likely to progress

Table 8.1 Reasons for benchmarking

Objectives	Without benchmarking	With benchmarking
Becoming competitive	• Internally focused • Evolutionary change	• Understanding of competition • Ideas from proven practices
Industry best practices	• Few solutions • Frantic catch-up activity	• Many options • Superior performance
Defining customer requirements	• Based on history or gut feeling • Perception	• Market reality • Objective evaluation
Establishing effective goals and objectives	• Lacking external focus • Reactive	• Credible, unarguable • Proactive
Developing true measures of productivity	• Pursuing pet projects • Strengths and weaknesses not understood • Route of least resistance	• Solving real problems • Understanding outputs • Based on industry best practices

through four focuses. Initially attention will be concentrated on competitive products or services, including, for example, design, development and operational features. This should develop into a focus on industry best practices and may include, for example, aspects of distribution or service. The real breakthrough is when the organization focuses on all aspects of the total business performance, across all functions and aspects, and addresses current and projected performance gaps. This should lead to the final focus on true continuous improvement.

At its simplest, competitive benchmarking, the most common form, requires every department to examine itself against the counterpart in the best competing companies. This includes a scrutiny of all aspects of their activities. Benchmarks which may be important for **customer satisfaction**, for example, might include:

- product or service consistency,
- correct and on-time delivery,
- speed of response or new product development,
- correct billing.

For **internal impact** the benchmarks may be:

- waste, rejects or errors,
- inventory levels/work in progress,
- costs of operation,
- staff turnover.

The task is to work out what has to be done to make improvements on the competition's performance in each of the chosen areas.

The purpose and practice of benchmarking

The purpose of benchmarking is predominantly to:

- **change** the perspectives of executives and managers
- **compare** business practices with those of world class organizations
- **challenge** current practices and processes
- **create** improved goals and practices for the organization.

As a managed process for change, benchmarking uses a disciplined structured approach to identify **what** needs to change, **how** it can be changed, and the **benefits** of the change. It also creates the desire for change in the first place.

Any process or practice that can be defined can be benchmarked but the focus should be on those which impact on customer satisfaction and/or business results – financial or non-financial.

At regular (say, weekly) meetings, managers discuss the results of the competitive benchmarking, and on a daily basis, departmental managers discuss quality problems with staff. One afternoon may be set aside for the benchmark meetings followed by a 'walkabout' when the manager observes directly the activities actually taking place and compares them mentally with the competitors' operations.

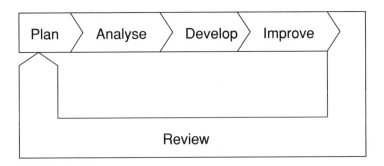

Figure 8.1 The five main stages of benchmarking

The process has five main stages which are all focused on trying to **measure** comparisons of competitiveness (Figure 8.1). The detail is as follows:

Plan   Select department(s) or process group(s) for benchmarking.
Identify best competitor, perhaps using customer feedback or industry observers.
Identify benchmarks – key performance indicators which will be used in the study.
Bring together the appropriate team to be involved.
Decide information and data collection methodology
(do not forget desk research!).
Prepare for any visits and interact with target organizations.
Use data collection methodology.

Analyse   Compare the organization and its 'competitors', using the benchmark data.

Catalogue the information and create a 'competency centre' which monitors capabilities.

Understand the 'enabling processes' as well as the performance measures.

Develop   Set new performance level objectives/standards.

Develop action plans to achieve goals and integrate into the organization including re-design of any processes identified as candidates.

Improve   Implement specific actions and integrate them into the business processes.

Review   Monitor the results and improvements.

Review the benchmarks and the ongoing relationship with the target organization.

Benchmarking is very important in the administration areas, since it continuously measures services and practices against the equivalent operation in the toughest direct competitors or organization renowned as leaders in the area, even if they are in the same organization. An example of quantitative benchmarks in absenteeism is given in Table 8.2.

Technologies and conditions vary between different industries and markets, but the basic concepts of measurement and benchmarking are of general validity. The objective should be to produce products and services which conform to the requirements of the customer in a never-ending improvement environment. The way to accomplish this is to use the PDCA continuous improvement cycle in all the operating departments – nobody should be exempt. Measurement and benchmarking are not separate sciences or unique theories of quality management, rather a strategic approach to getting the best out of people, processes, products, plant, and programmes.

Gaining management support and direction

The benchmarking effort needs senior management support and this may be obtained by preparing and presenting an executive briefing. This should relate

Table 8.2 Quantitative benchmarking in absenteeism

Organization's absence level	Productivity opportunity
Under 3%	This level matches an aggresive benchmark which has been achieved in 'excellent' organizations.
3–4%	This level may be viewed within the organization as a good absence performance – represents a moderate productivity opportunity improvement.
5–8%	This level is tolerated by many organizations – represents a major improvement opportunity.
9–10%	This level indicates that a serious absence problem exists.
Over 10%	This level of absence is totally unacceptable.

to the corporate goals and provide examples of success from suitable research. An 'executive champion' should be chosen if possible and a benchmarking steering committee formed.

The role of the steering committee is to:

- guide the selection of benchmarking areas
- provide contacts with external organizations
- support the 'executive champion'
- promote benchmarking amongst the senior management

The group can also facilitate the integration of benchmarking into 'difficult' areas of the business.

Are you ready for benchmarking?

A simple self-assessment pro forma is provided in Table 8.3 to help you determine:

- how well you understand your processes
- how much you listen to your customers
- how committed your senior team is.

Table 8.3   Are you ready for benchmarking?

Study the statements below and tick one box for each to reflect the level to which the statement is true for your organization.

	Most	Some	Few	None
Processes have been documented with measures to understand performance.	☐	☐	☐	☐
Employees understand the processes that are related to their own work.	☐	☐	☐	☐
Direct customer interactions, feedback or studies about customers influence decisions about products and services.	☐	☐	☐	☐
Problems are solved by teams.	☐	☐	☐	☐
Employees demonstrate by words and deeds that they understand the organization's mission, vision and values.	☐	☐	☐	☐
Senior executives sponsor and actively support quality improvement projects.	☐	☐	☐	☐
The organization demonstrates by words and by deeds that continuous improvement is part the culture.	☐	☐	☐	☐
Commitment to change is articulated in the organization's strategic plan.	☐	☐	☐	☐
Add the columns:.	☐	☐	☐	☐
	× 6 =	× 4 =	× 2 =	Zero
Multiply by the factor	☐	☐	☐	☐

What is the grand total? ☐

What score did you obtain in Table 8.3?

    32–48   You are ready for benchmarking!

    16–31   You need some preparation!

    0–15   You need some help!

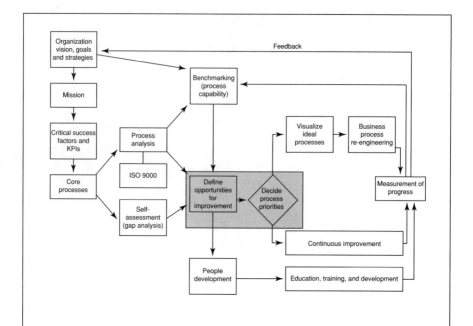

Organization
vision, goals
and strategies

Feedback

Benchmarking
(process
capability)

Mission

Critical success
factors and
KPIs

Process
analysis

Visualize
ideal
processes

Business
process
re-engineering

Core
processes

ISO 9000

Define
opportunities
for
improvement

Decide
process
priorities

Measurement of
progress

Self-
assessment
(gap analysis)

Continuous improvement

People
development

Education, training, and development

Defining improvement opportunities and prioritizing ____

Chapter 9

Defining improvement opportunities and prioritizing _____

Key points

A set of simple tools is needed to assist in defining opportunities for improvement and prioritizing the actions needed. The effective use of the tools requires the commitment and involvement of the people who work on the processes. This in turn needs management support and the provision of training.

This chapter explains some basic tools for defining opportunities and prioritizing, such as: process flow charting (Chapter 5), check/tally charts, histograms, scatter diagrams, stratification, Pareto analysis, cause and effect analysis and brainstorming (also CEDAC, NGT and the five whys), force-field analysis, and simple control charts.

Also discussed is the use of some additional tools in defining and prioritizing the improvement of processes. These are systems and documentation methods for identifying objectives and intermediate steps in detail: affinity diagrams, interrelationship digraphs, systems flow/tree diagrams, matrix diagrams and the process decision programme chart (PDPC).

These tools are interrelated and their promotion and use should lead to efficient and effective process improvements in less time. They work best when people from all parts of an organization are involved. Some of the tools can be used in general problem solving.

Failure mode, effect and criticality analysis (FMECA) is discussed briefly as it is the study of potential system or process failures, their effects and criticality. Moments of truth (MoT), a similar concept to FMEA, is the moment in time when a customer first comes into contact with an organization, leading to a judgement, and this is also covered. FMECA and MoT are excellent methods of identifying opportunities for improvement.

The 'tool kit' offered is a compilation and modification of techniques that have been around for a long time. These should be used together to make better informed decisions about process improvement and/or re-design.

A systematic approach

An organization may identify opportunities for improvement in a number of ways, perhaps through process analysis (Chapter 5), benchmarking (Chapter 8), or the use of self-assessment against an established framework (Chapter 7). The use of the simple tools to be covered in this chapter further aid understanding and identification and assist in the prioritization for action. The total organization framework shows, however, that a distinction needs to be made between those processes which run pretty well – to be subjected to a regime of continuous improvement (Chapter 10), and those which are very poor and are in need of a complete re-visioning, re-design or 're-engineering' activity (Chapter 11.) This chapter attempts to provide help in these vital defining and prioritizing activities for improvement.

In the never-ending quest for improvement of processes, information will form the basis for understanding, decisions and actions, and a thorough data gathering, recording and presentation system is essential.

In addition to the basic elements of a management system which provides a framework for recording, there exists a set of 'tools' which can be used to interpret fully and derive maximum use of data. The simple methods listed below will offer any organization means of collecting, presenting and analysing most of its data:

- Process flow charting   – what is done?

- Check sheets/tally charts   – how often is it done?

- Histograms   – what do overall variations look like?

- Scatter diagrams   – what are the relationships between factors?

- Stratification   – how is the data made up?

- Pareto analysis   – which are the big problems?

- Cause and effect analysis and brainstorming (including CEDAC, NGT and the five whys)   – what causes the problems?

- Force-field analysis   – what will obstruct or help the change or solution?

- Control charts   – which variations to control and how?

The effective use of the tools requires the involvement of the people who actually work on the processes. Their commitment to this will be possible only if they are assured that management cares about improvement. The latter must show they are serious by establishing a systematic approach and providing the training and implementation support required.

Improvements cannot be achieved without specific opportunities being identified or recognized, a focus on which leads to the creation of teams, whose membership is determined by their involvement in and detailed knowledge of the process, and their ability to take improvement action. The teams must then be provided with good leadership and the right tools to tackle the job.

The systematic approach mapped out in Figure 9.1 should lead to the use of factual information, collected and presented using proven techniques, to open good channels of communication.

By using reliable methods, creating a favourable environment for team based process investigation, and continuing to improve using systematic techniques, the never-ending improvement helix (see Chapter 2) will be engaged. This approach demands the real-time management of data, and actions on processes and inputs, not outputs. It will require a change in the language of many organizations from percentage defects, percentage 'prime' product, and number of errors, to **process capability**. The driving force for this will be the need for better internal and external customer satisfaction levels which will lead to the key improvement question, 'Could we do the job better?'

Some basic tools and techniques

Understanding processes so that they can be improved using the systematic approach requires knowledge of a simple kit of tools or techniques. What follows is a brief description of each technique, but a full description and further examples of some of them may be found in *Statistical Process Control*, 4th Edition, by John S. Oakland, Butterworth-Heinemann, Oxford 1999.

Process flow charting/mapping

The use of these techniques, which were described more fully in Chapter 5, ensures an understanding of the inputs and flow of the process. Without that understanding, it is not possible to draw the correct flow chart of the process. In flow charting and mapping it is important to remember that in all but the smallest tasks, no single person is able to complete it without help from others who work in the process. This makes flow charting/mapping powerful team-forming exercises.

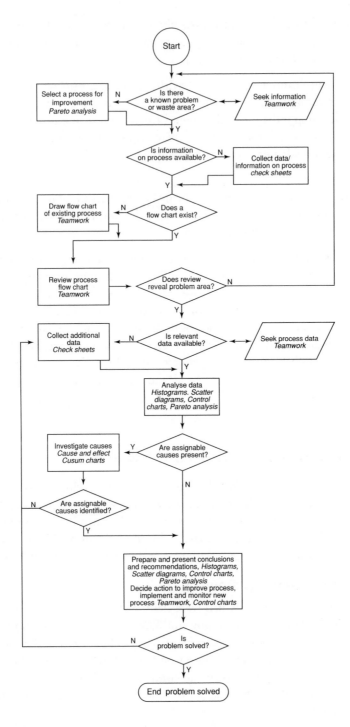

Figure 9.1   Strategy for process improvement

Check sheets or tally charts

A check sheet is a tool for data gathering and a logical point to start in most process control or problem-solving efforts. It is particularly useful for recording direct observations and helping to gather in facts rather than opinions about the process.

In the recording process, it is essential to understand the difference between data and numbers. Data is information, including numerical, that is useful in solving problems, or that provides knowledge about the state of a process. Numbers alone often represent meaningless measurements or counts which tend to confuse rather than to enlighten. Numerical data on quality will arise either from:

1   Counting, or

2   Measurement.

Data which arises from counting can occur only at definite points or in 'discrete' jumps. There can be only 0, 1, 2, etc., errors in an invoice page, there cannot be 2.46 errors. The number of pens which fail to write properly give rise to discrete data which are called **attributes**. As there is only a two-way classification to consider right or wrong, present or not present, attributes give rise to counted data, which necessarily varies in jumps.

Data which arises from measurement can occur anywhere at all on a continuous scale and is called **variable** data. The weight of a capsule, the diameter of a piston, the tensile strength of a piece of rod, the time taken to process an insurance claim, are all variables, the measurement of which produces continuous data.

The use of simple check sheets or tally charts aids the collection of data of the right type, in the right form, at the right time. The objectives of the data collection will determine the design of the record sheet used, which in turn give rise to frequency distributions.

Histograms

Histograms show, in a very clear pictorial way, the frequency with which a certain value or group of values occurs. They can be used to display both attribute and variables data, and are an effective means of communicating directly to the people who operate the process, the results of their efforts. Data gathered on a tally chart is often drawn as a histogram, e.g. Figure 9.2.

Scatter diagrams

Depending on the technology, these are frequently useful to establish the asso-

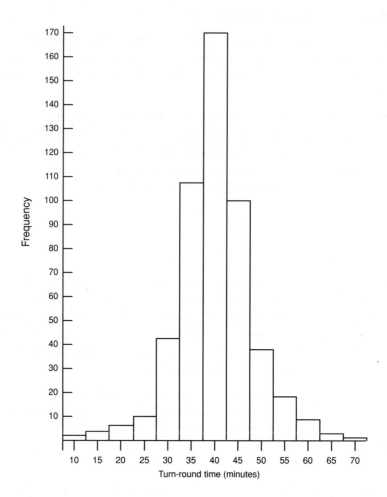

Figure 9.2 Frequency distribution for truck turn-round times (histogram)

ciation, if any, between two parameters or factors. A technique to begin such an analysis is a simple X–Y plot of the two sets of data. The resulting grouping of points on scatter diagrams (e.g. Figure 9.3) will reveal whether or not a strong or weak, positive or negative correlation exists between the parameters. The diagrams are simple to construct and easy to interpret, and the absence of correlation can be as revealing as finding that a relationship exists.

Stratification

Stratification is simply dividing a set of data into meaningful groups, and can be used to great effect in combination with other techniques, including his-

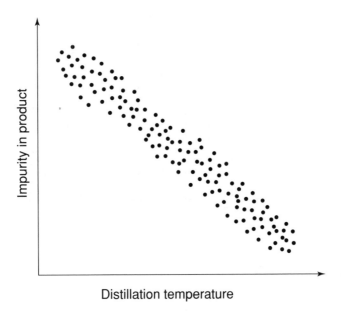

Figure 9.3   Scatter diagram showing a negative correlation between two variables

tograms and scatter diagrams. If, for example, three shift teams are responsible for the output described by the histogram (a) in Figure 9.4, 'stratifying' the data into the shift groups might produce histograms (b), (c) and (d) and indicate process adjustments which were taking place at shift change-overs.

Figure 9.5 shows the scatter diagram relationship between advertising investment and revenue generated for all products. In diagram (a) all the data are plotted, and there seems to be no correlation. But if the data are stratified according to product, a correlation is seen to exist. Of course, the reverse may be true, so the data should be kept together and plotted in different colours or symbols to ensure all possible interpretations are retained.

Pareto analysis

If the symptoms or causes of defective output or some other 'effect' are identified and recorded, it will be possible to determine what percentage can be attributed to any cause, and the probable result will be that the bulk (typically 80 per cent) of the errors, waste, or 'effects' derive from a few of the causes (typically 20 per cent). For example, Figure 9.6 shows a **ranked frequency distribution** of incidents in the distribution of a certain product. To improve the performance of the distribution process, therefore, the major incidents (broken bags/drums, truck scheduling, temperature problems) should be

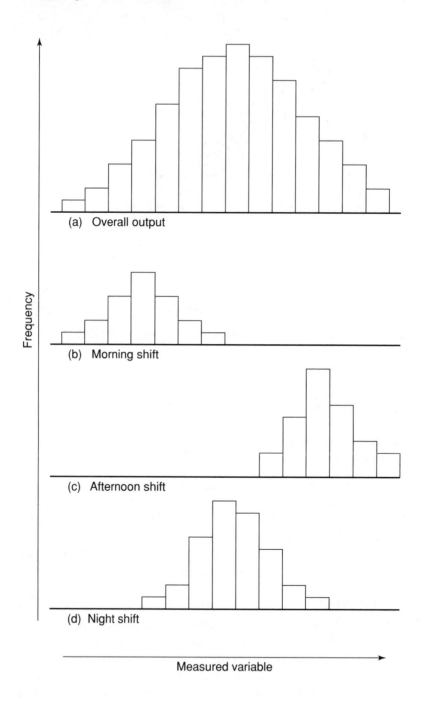

(a) Overall output

(b) Morning shift

(c) Afternoon shift

(d) Night shift

Frequency

Measured variable

Figure 9.4   Stratification of data into shift teams

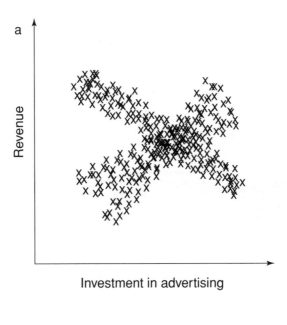

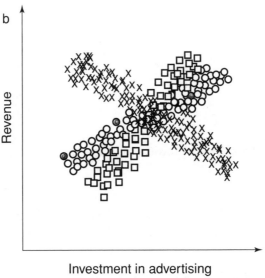

Figure 9.5 Scatter diagram of investment in advertising versus revenue: (a) without stratification, (b) with stratification

tackled first. An analysis of data to identify the major problems is known as **Pareto analysis**, after the Italian economist who realized that approximately 90 per cent of the wealth in his country was owned by approximately 10 per

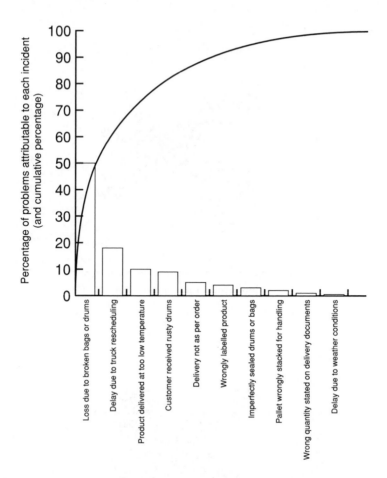

Figure 9.6   Incidents in the distribution of a chemical product

cent of the people. Without an analysis of this sort, it is far too easy to devote resources to addressing one symptom only because its cause seems immediately apparent.

Cause and effect analysis and brainstorming

A useful way of mapping the inputs which affect quality is the **cause and effect diagram,** also known as the Ishikawa diagram (after its originator) or the fishbone diagram (after its appearance – Figure 9.7). The effect or incident being investigated is shown at the end of a horizontal arrow. Potential causes are then shown as labelled arrows entering the main cause arrow. Each arrow may have other arrows entering it as the principal factors or causes are reduced to their sub-causes, and sub-sub-causes by **brainstorming.**

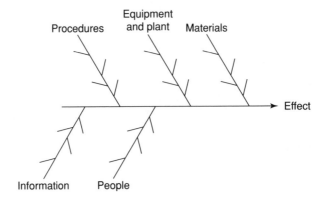

Figure 9.7   The cause and effect, Ishikawa or fishbone diagram

Brainstorming is a technique which is used to generate a large number of ideas quickly and may be used in a variety of situations. Each member of a group, in turn, may be invited to put forward ideas concerning a problem under consideration. Wild ideas are safe to offer, as criticism or ridicule is not permitted during a brainstorming session. The people taking part do so with equal status to ensure this. The main objective is to create an atmosphere of enthusiasm and originality. All ideas offered are recorded for subsequent analysis. The process is continued until all the conceivable causes have been included. The proportion of non-conforming output attributable to each cause, for example, is then measured or estimated, and a simple Pareto analysis identifies the major causes which are most worth investigating. A useful variant on the technique is negative brainstorming and cause/effect analysis. Here the group brainstorms all the things which would need to be done to ensure a negative outcome. Having identified the potential 'road blocks', it is easier to dismantle them.

CEDAC

A variation on the cause and effect approach, which was developed at Sumitomo Electric and now is claimed to be used by major Japanese corporations across the world, is the Cause and Effect Diagram with Addition of Cards (CEDAC).

The effect side of a CEDAC chart is a quantified description of the problem, with an agreed and visually quantified target, and continually updated results on the progress of achieving it. The cause side of the CEDAC chart uses two different coloured cards for writing **facts** and **ideas**. This ensures that the facts are collected and organized, before solutions are devised.

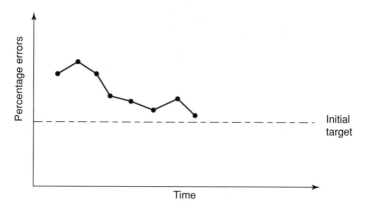

Figure 9.8 The effects side of a CEDAC chart showing the target and quantified improvements

The basic diagram for CEDAC has the classic fishbone appearance, and this is drawn on a large piece of paper, with the effect on the right and cause on the left. A project leader is chosen to be in charge of the CEDAC team and sets the improvement target. A method of measuring and plotting the results on the effects side of the chart is devised so that a visual display of the target and the quantified improvements are provided (for example, see Figure 9.8).

The **facts** are gathered and placed on the left of the spines on the cause side of the CEDAC chart (Figure 9.9). The people in the team submitting the fact cards are required to initial them. Improvement **ideas** cards are then generated and placed on the right of the cause spines. The ideas are then selected and evaluated for substance and practicality. The test results are recorded on the effect side of the chart.

Successful improvement ideas are incorporated into the new process. CEDAC is another systematic approach to marshalling the creative resources and knowledge of the people involved in processes. When they own and can measure the problems, they will find the solutions.

Nominal Group Technique (NGT)

The Nominal Group Technique (NGT) is a particular form of team brain-storming used to prevent domination by particular individuals. It has specific application for multi-level, multi-disciplined teams where communication boundaries are potentially problematic.

In NGT, a carefully prepared written statement of the improvement area or problem to be tackled is read out by the facilitator (F). Clarification is

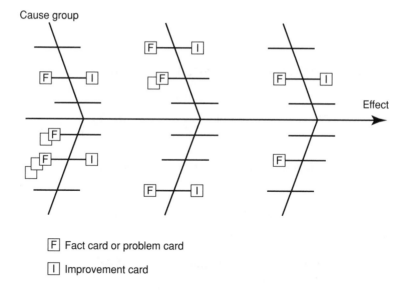

Figure 9.9 The CEDAC diagram with fact and improvement cards

obtained by questions and answers, and then the individual participants (P) are asked to re-state the problem in their own words. The group then discuss the problem until its formulation can be satisfactorily expressed by the team (T). The method is set out in Figure 9.10.

About fifteen minutes is then allowed for **silent idea generation** in which the participants think quietly about their responses and suggestions for the problem statement. In a **round-robin** process, each member of the team is asked in turn to put forward one suggestion or idea. Additional ideas which are stimulated by others' suggestions are also collected on flip charts. As with any form of brainstorming, no criticism is allowed by team members or facilitator. The round-robin continues until no more ideas/suggestions are forthcoming.

The **clarification** stage begins with the facilitator going through each suggestion in turn for explanation, as necessary. Long detailed discussions and the premature working of issues by the team must be controlled by the facilitator. After any deletions and/or rephrasing, **selection and ranking** takes place using cards issued to all team members. On these the members are asked to write, in silence, their top five or six ideas with rankings. The facilitator collects the cards and transfers the results onto a new flip chart. The ranking scores are summed to give a **final ranking.** The result is a set of ranked ideas which are close to a team consensus view, obtained without domination by one or two individuals.

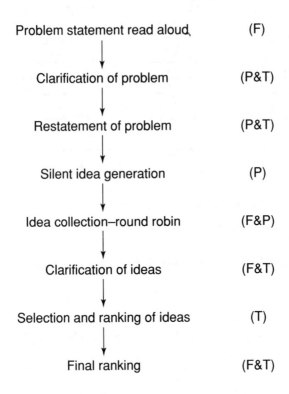

Figure 9.10 Nominal group technique (NGT). F, facilitator; P, participants; T, team

The five whys

This is a systematic questioning approach to ensure that the root causes of problems are sought out. It consists merely of asking 'why?' several times in succession. The originators of the method, Toyota, suggested that 'why' be asked successively at least five times to ensure root cause is truly established. It can be very helpful in distinguishing between processes which require improvement and those as candidates for re-engineering.

Force-field analysis

Force-field analysis is a technique used to identify the forces that either obstruct or help a change that needs to be made. It is similar to negative brain-storming and cause/effect analysis and helps to plan how to overcome the barriers to change or improvement. It may also provide a measure of difficulty in achieving the change.

The process begins with a team describing the desired change or improvement, and defining the objectives or solution. Having prepared the basic force-field diagram, the favourable/positive/driving forces, and the unfavourable/negative/restraining forces are identified using brainstorming. These are placed in opposition on the diagram and, if possible, rated for their potential influence on the ease of implementation. The results are evaluated and an action plan prepared to overcome some of the restraining forces, and increase the driving forces. Figure 9.11 shows a force-field diagram produced by a senior management team considering the implementation of 'business excellence' in their organization.

Control charts

A control chart is a form of traffic signal, the operation of which is based on evidence from the small samples taken at random during a process. A green light is given when the process should be allowed to run. All too often, processes are 'adjusted' on the basis of a single measurement, check or inspection, a practice which can make a process much more variable than it is already. The equivalent of an amber light appears when trouble is possibly imminent. The red light shows that there is practically no doubt that the process has changed in some way and that it must be investigated and corrected to prevent production of defective material or information. Clearly, such a scheme can be introduced only when the process is 'in control'. Since

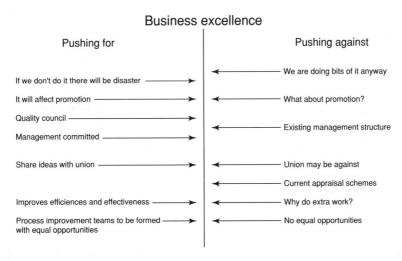

Figure 9.11   Force-field analysis

samples taken are usually small, there are risks of errors, but these are small, calculated risks and not blind ones. The risk calculations are based on various frequency distributions.

These charts should be made easy to understand and interpret and they can become, with experience, sensitive diagnostic tools which can be used by operating staff and first-line supervision to prevent errors or defective output being produced. Time and effort spent to explain the working of the charts to all concerned is never wasted.

The most frequently used control charts are simple run charts where the data is plotted on a graph against time or sample number. There are different types of control charts for variables and attribute data: for variables, mean ($\bar{X}$) and range (R) charts are used together; number defective or *np* charts and proportion defective or *p* charts are the most common ones for attributes. Other charts found in use are moving average and range charts, number of defects (*c* and *u*) charts, and cumulative sum (cusum) charts. The latter offer very powerful management tools for the detection of trends or changes in attributes and variable data.

The cusum chart is a graph which takes a little longer to draw than the conventional control chart, but which gives a lot more information. It is particu-

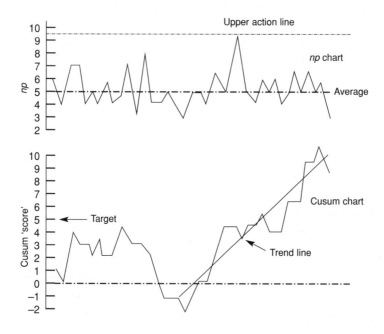

Figure 9.12   Comparison of cusum and *np* charts for the same data

larly useful for plotting the evolution of processes because it presents data in a way that enables the eye to separate true changes from a background of random variation. Cusum charts can detect small changes in data very quickly and may be used for the control of variables and attributes. In essence, a reference or 'target value' is subtracted from each successive sample observation and the result accumulated. Values of this cumulative sum are plotted and 'trend lines' may be drawn on the resulting graphs. If this is approximately horizontal, the value of the variable is about the same as the target value. An overall slope downwards shows a value less than the target, and if the slope is upwards it is greater. The technique is very useful, for example, in comparing sales forecast with actual sales figures.

Figure 9.12 shows a comparison of an ordinary run chart and a cusum chart which have been plotted using the same data – errors in samples of 100 invoices. The change, which is immediately obvious on the cusum chart, is difficult to detect on the conventional control chart.

The range of type and use of control charts is now very wide and within the present text it is not possible to indicate more than the basic principles underlying such charts. This is done in the book *Statistical Process Control*, by Oakland which provides a comprehensive text on the subject.

Some additional tools for improvement, identification and prioritization

There are some additional systems and documentation methods which can be used to identify and prioritise improvement opportunities in some detail, including:

- affinity diagram
- inter-relationship digraph
- systems flow/tree diagram
- matrix diagram
- process decision programme chart

Affinity diagram

This is used to gather large amounts of language data (ideas, issues, opinions) and organize it into groupings based on the natural relationship between the items. In other words, it is a form of brainstorming. One of the obstacles often encountered in the quest for improvement is past success or failure. It is assumed that what worked or failed in the past will continue to do so in the future. Although the lessons of the past should not be ignored, unvarying patterns of thought which can limit progress should not be enforced. This is especially true in potential process re-engineering situations where **new** logical patterns should always be explored.

The affinity diagram, like other brainstorming methods, is part of the creative process. It can be used to generate ideas and categories that can be used later with more strict, logic based tools. This tool may be used to 'map the geography' of a process or issue when:

- facts or thoughts are in chaos and the issues are too complex to easily define,

- breakthroughs in traditional concepts are needed to replace old solutions and to expand a team's thinking,

- support for a solution is essential for successful implementation.

The affinity diagram is not recommended when a problem is simple or requires a very quick solution. It is a compilation of a maximum number of ideas under a limited number of major headings (see for example Figure 9.13). This data can then be used with other tools to logically define areas for attack. One of these tools is the inter-relationship digraph.

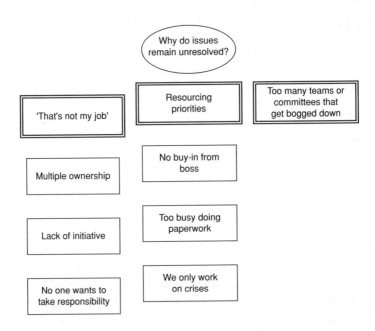

Figure 9.13 Example of an affinity diagram

Inter-relationship digraph

This tool is designed to take a central idea, issue or problem and map out the logical or sequential links among related factors. While this still requires a very creative process, the inter-relationship digraph begins to draw the logical connections that surface in the affinity diagram.

In process design, planning, and problem solving, it is obviously not enough to just create an explosion of ideas. The affinity diagram method allows some organized creative patterns to emerge but the inter-relationship digraph lets **logical** patterns become apparent. This tool starts, therefore, from a central concept, leads to the generation of large quantities of ideas and finally the delineation of observed patterns. Like the affinity diagram, the interrelationship digraph allows those unanticipated ideas and connections to rise to the surface.

The technique is adaptable to both specific operational issues and general organizational questions. For example, a classic use of this tool at Toyota focused on all of the factors involved in the establishment of a 'billboard system' as part of their JIT programme. It has also been used to get top management support for process management and improvement.

In summary, the inter-relationship digraph should be used when:

(a) An issue is sufficiently complex that the interrelationship between ideas is difficult to determine.

(b) The correct sequencing of management actions is critical.

(c) There is a feeling or suspicion that the problem under discussion is only a symptom.

(d) There is ample time to complete the required reiterative process and define cause and effect.

The inter-relationship digraph can be used by itself, or it can be used after the affinity diagram, using data from the previous effort as input. Figure 9.14 gives an example of a simple inter-relationship digraph.

Systems flow/tree diagram

The systems flow/tree diagram (usually referred to as tree diagram) is used to systematically map out the full range of activities that must be accomplished in order to reach a desired goal. It may also be used to identify all of the factors contributing to a problem under consideration. Major factors identified using an interrelationship digraph can be used as inputs for the tree diagram. One of the strengths of this method is that it forces the user to examine the logical

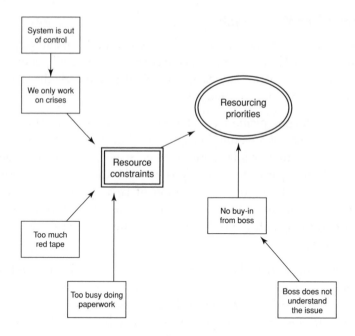

Figure 9.14 Example of the inter-relationship digraph

and chronological links between tasks. This assists in avoiding a natural tendency to jump directly from goal or problem statement to solution (ready...fire...aim!).

The tree diagram is indispensable when a thorough understanding of what needs to be accomplished is required, together with how it is to be achieved, and the relationships between these goals and methodologies. It has been found to be most helpful in situations when:

(a) very ill-defined needs must be translated into operational characteristics, and to identify which characteristics can presently be controlled,

(b) all the possible causes of a problem need to be explored. This use is closest to the cause and effect diagram or fishbone chart,

(c) identifying the first task that must be accomplished when aiming for a broad organizational goal,

(d) the issue under question has sufficient complexity and time available for solution.

Depending on the type of issue being addressed, the tree diagram will be similar to either a cause and effect diagram, or a flow chart, although it might be easier to interpret because of its clear linear layout. If a problem is being

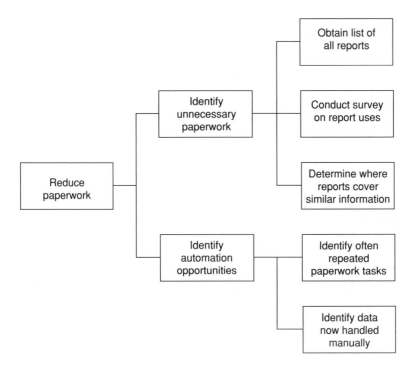

Figure 9.15 An example of the tree diagram

considered, each branch of the tree diagram will be similar to a cause and effect diagram. If a general objective is being considered, each branch may represent chronological activities, in which case the diagram will be similar to a flow chart. An example is shown in Figure 9.15.

Matrix diagram

The purpose of the matrix diagram is to outline the interrelationships and correlations between tasks, functions or characteristics and to show their relative importance. There are many versions of the matrix diagram, but the most widely used are the simple L-shaped matrix and the T matrix.

(i) L-shaped matrix diagram

This is the most basic form of matrix diagram. In the L shape two interrelated groups of items are presented in line and row format. It is a simple two-dimensional representation that shows the intersection of related pairs of items as shown in Figure 9.16. It can be used to display relationships between items in all operational areas, including administration, manufacturing, personnel, R&D etc., to identify all the organizational tasks that need to be accomplished and how they should be allocated to individuals.

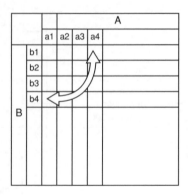

Figure 9.16   L-shaped matrix

In one application, customer demands (the 'whats') are analysed with respect to substitute quality characteristics (the 'hows') (Figure 9.17). Correlations between the two are categorized as strong, moderate and possible. The customer demands shown on the left of the matrix are determined in co-operation with the customer. This effort requires a kind of a verbal 'ping-pong' with the customer to be truly effective: ask the customer what he wants, write it down, show it to him and ask him if that is what he meant, then revise and repeat the process as necessary. This should be done in a joint meeting with the customer, if at all possible. It is often of value to use a tree diagram to give structure to this effort.

The right side of the chart is often used to compare current performance to competitors' performance, company plan and potential sales points with ref-

Substitute quality characteristics

	MFR	Ash	Importance	Current	Best competitior	Plan	IR	SP	RQW
No film breaks	◯ 17	▲ 6	4	4	4	4	1	◯	5.6
High rates	◉ 23		3	3	4	4	1.3		4.6
Low gauge variability	◉ 37	▲ 7	4	3	4	4	1.3	◯	7.3

Customer demands

◉ Strong correlation   IR   Improvement ratio

◯ Some correlation   SP   Sales point

▲ Possible correlation   RQW   Relative quality weight

Figure 9.17   An example of the matrix diagram

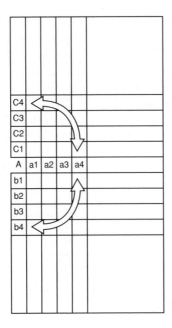

Figure 9.18   T-shaped matrix

erence to the customer demands. Weights are given to these items to obtain a 'relative quality weight'. This can be used to identify the key customer demands. The relative quality weight is then used with the correlations identified on the matrix to determine the key quality characteristics.

(ii) T-shaped matrix diagram

The T-shaped matrix is nothing more than the combination of two L-shaped matrix diagrams. As can be seen in Figure 9.18, it is based on the premise that two separate sets of items are related to a third set. Therefore, A items are somehow related to both B and C items.

Figure 9.19 shows one application. In this case, it shows the relationship between a set of courses in a curriculum and two important sets of considerations: who should do the training for each course and which would be the most appropriate functions to attend each of the courses.

It has also been widely used to develop new materials by simultaneously relating different alternative materials to two sets of desirable properties.

There are other matrices that deal with ideas such as product or service function, cost, failure modes, capabilities, etc., and there are at least 40 different types of matrix diagrams available.

Who trains?

	SPC	7 old tools	7 new tools	Reliability	Design review	QC basics	QCC facilitator	Diagnostic tools	Problem solving	Communication skills	Organize for quality	Design of experiments	Company mission	Quality planning	Just in time	New superv. training	Company TQM system	Group dynamics skills	SPC course/execs
Human resources dept																			
Managers																			
Operators*																			
Consultants																			
Production operator																			
Craft foreman																			
GLSPC coordinator																			
Plant SPC coordinator																			
University																			
Technology specialists																			
Engineers																			

* Need to tailor to groups

X= Full
O= Overview

Courses

Who attends?

	SPC	7 old tools	7 new tools	Reliability	Design review	QC basics	QCC facilitator	Diagnostic tools	Problem solving	Communication skills	Organize for quality	Design of experiments	Company mission	Quality planning	Just in time	New superv. training	Company TQM system	Group dynamics skills	SPC course/execs
Executives																			
Top management																			
Middle management																			
Production supervisors																			
Supervisor functional																			
Staff																			
Marketing																			
Sales																			
Engineers																			
Clerical																			
Production worker																			
Quality professional																			
Project team																			
Employee involvement																			
Suppliers																			
Maintenance																			

Figure 9.19 T-matrix diagram on company-wide training

Process decision programme chart

A process decision programme chart (PDPC) is used to map out each event and contingency that can occur when progressing from a problem statement to its solution. The PDPC is used to anticipate the unexpected and plan for it. The PDPC is related to a failure mode and effect analysis and its structure is similar to that of a tree diagram. An example of the PDPC is shown in Figure 9.20.

The PDPC is very simply an attempt to be proactive in the analysis of failure and to construct, on paper, a 'dry run' of the process so that the 'check' part of the improvement cycle can be defined in advance. PDPC is likely to enjoy widespread use in defining and prioritizing improvement opportunities.

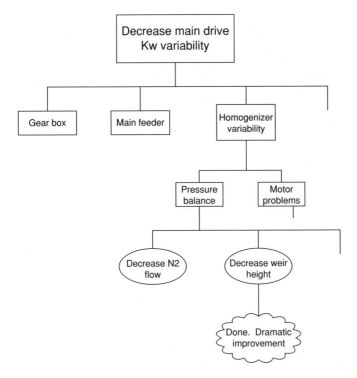

Figure 9.20 Process decision programme chart

Failure mode, effect and criticality analysis (FMECA)

It is possible to analyse processes to determine possible modes of failure and their effects on the performance of the product or operation of the process or service system. Failure mode and effect analysis (FMEA) is the study of potential failures to determine their effects. If the results of an FMEA are ranked in order of seriousness, then the word **criticality** is added to give FMECA. The primary objective of a FMECA is to determine the features of process design or operation which are critical to the various modes of failure, in order to reduce failure. It uses all the available experience and expertise, from marketing, design, technology, purchasing, production/operation distribution, service, etc., to identify the importance levels or criticality of potential problems and stimulate action which will reduce these levels. FMECA should be a major consideration at the design stage of a product or service.

The elements of a complete FMECA are:

(i) **Failure mode.** The anticipated conditions of operation are used as the background to study the most probable failure mode, location and mechanism of the process or system and its components.

(ii) **Failure effects**. The potential failures are studied to determine their probable effects on the performance of the whole process and the effects of the various components on each other.

(iii) **Failure criticality**. The potential failures in the various parts of the process or system are examined to determine the severity of each failure effect in terms of lowering of performance, safety hazard, total loss of function, etc.

FMECA may be applied at any stage of process design, development, or operation but since its main aim is to prevent failure, it is most suitably applied at the design stage to identify and eliminate causes. With more complex processes or systems, it may be appropriate to consider these as sub-systems, each one being the subject of a separate FMECA.

Special FMECA pro formas are available which set out the steps of the analysis as follows:

1   Identify the process or system components, or process function.

2   List all possible failure modes of each component.

3   Set down the effects that each mode of failure would have on the overall function of the process or system.

4   List all the possible causes of each failure mode.

5   Assess numerically the failure modes on a scale from 1 to 10. Experience and reliability data should be used, together with judgement, to determine the values, on a scale 1–10, for:

P   the probability of each failure mode occurring (1 = low, 10 = high).
S   the seriousness or criticality of the failure (1 = low, 10 = high).
D   the difficulty of detecting the failure before the product or service is used by the consumer (1 = easy, 10 = very difficult).

Value	1	2	3	4	5	6	7	8	9	10
P	low chance of occurrence---------------------almost certain to occur									
S	not serious, minor nuisance---------------total failure, safety hazard									
D	Easily detected------------------------------------unlikely to be detected									

Calculate the product of the ratings, $C = P \times S \times D$, known as the criticality index or risk priority number (RPN) for each failure mode. This indicates the relative priority of each mode.

6   Indicate briefly the corrective action required and, if possible, which department or person is responsible and the expected completion date.

When the criticality index has been calculated, the failures may be ranked accordingly. It is usually advisable, therefore, to determine the value of C for each failure mode before completing the last columns. In this way, the action required against each item can be judged in the light of the ranked severity and the resources available.

Moments of truth (MoT) is a concept which has much in common with FMEA. A MoT is the moment in time when a customer first comes into contact with the people, processes or systems of an organization, which leads to the customer making a judgement about the quality of the organization's services or products.

In MoT analysis the points of potential dissatisfaction are identified proactively, and it begins with the assembly of process flow chart-type diagrams. Every small step taken by a customer in his/her dealings with the organization's people, processes, or services is recorded. It may be difficult or impossible to identify all the MoTs but the systematic approach should lead to a minimization of the number and the severity of unexpected failures, and this provides the link with FMEA.

Prioritizing for improvement and breakthrough

What has been described in this chapter is a tool kit for improvement identification and prioritization. It is a compilation and modification of some tools that have been around for a long time that are meant to be used together in making better informed decisions about processes and their improvement or redesign. In addition to the structure that the tools provide, the co-operation between functions or departments that is required in their use will help break down barriers within organizations and the tools work best when representatives from all parts of an organization are involved.

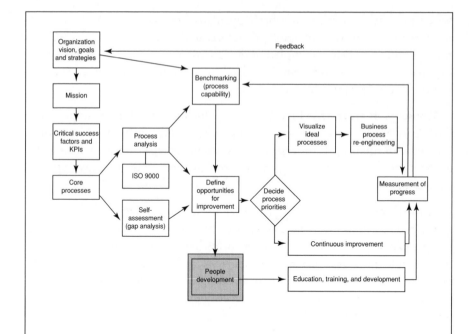

People – their development and teamwork _____

Chapter 10

People – their development and teamwork _____

Key points

The only point at which true responsibility for performance can lie is with the person or group actually doing a job and this chapter shows how organizations can make more effective use of their people through **performance management.**

Managers are in control only when their subordinates can exercise self-control. The process of performance management then consists of clarifying responsibilities, developing performance indicators and objectives, and preparing action plans.

The only efficient way to tackle process improvement or complex problems is through **teamwork**, which allows individuals and organizations to grow. Of course, employees will not engage continual improvement without commitment from the top, a 'climate' for improvement and an effective mechanism for capturing individual contributions. Teamwork must be driven by a strategy, have a structure, and be implemented thoughtfully and effectively.

John Adair's model for teamwork, which addresses the needs of the task, the team, and the individuals in the team, in the form of three overlapping circles, is used for the main structure of this chapter.

When teams are put together they pass through Tuckman's forming (awareness), storming (conflict), norming (co-operation), and performing (productivity) stages of development.

The Myers-Briggs Type Indicator (MBTI) and Schutz's FIRO-B (Fundamental Interpersonal Relationship Orientation-Behaviour) instrument are introduced as techniques for helping individuals understand and value each other for their differences as well as their similarities. It will also form the basis for thinking on culture change – a new 'openness cycle' is presented, based on inclusion, control, and openness, i.e., the FIRO-B model.

For any of the personality or teamwork models to succeed, groups need to be facilitated through the 'five A' stages of: becoming **aware**, needing to **accept**, **adopt** and **adapt** in order to **act.**

In thinking about achieving total organizational excellence, it may be useful to ask first if the managers have the necessary authority, capability and time to carry through the required changes. A disciplined and systematic approach to continuous improvement may be established in a **steering group** – the members of which are the senior management team, reporting to whom are the process teams or any site steering committees, which in turn control the improvement teams.

Finally team performance is linked to the culture of the organization and, in particular, the senior management's willingness to embrace the principles of teamworking themselves and improve the ways in which each pair of individuals work together.

Responsibilities and performance management

The establishment of positive business objectives within an organization must be accompanied by the clear allocation of responsibilities within the people structure. It is generally accepted that the primary operational responsibility for ensuring that translation of goals takes place in the organization must rest with management, but true goal translation is impossible without the full co-operation and commitment of all employees. If they are to accept their full share of responsibility they must be able to participate fully in the making and monitoring of arrangements for achieving the requirements. Some organizations have arrangements whereby people – say in a particular unit – meet periodically for discussions. This 'total involvement' approach stresses the need for the participation of every individual employee.

The process of performance management

Managers are in control only when they have created a system and climate in which their subordinates can exercise self-control. Mechanisms may then be created to provide clear performance standards in all areas, backed by appropriate job descriptions and training, to ensure those standards are achieved. The process of performance management consists then of:

1 Clarifying responsibilities.

2 Developing performance indicators and objectives.

3 Preparing action plans.

1 Clarifying responsibilities

If job descriptions have been written for the organization, they may serve as a starting point for clarifying each individual's role. It should be emphasized, however, that these need to be updated and reviewed with each subordinate to ensure their relevancy. The format of job descriptions is not of critical importance although they must be standardized for a particular organization. They should contain a statement of the overall purpose, reporting relationships, responsibility and priorities. Agreement must be reached on the priorities – frequently managers or directors expect the major emphasis to be on activities 2, 5, 6, and 7, when the subordinate perceives the critical areas to be 1, 3, 4, and 8.

Organizational problems, which require role clarification, will not be resolved by the introduction of job descriptions alone. Some of the other factors, which prevent people from functioning smoothly, are:

(i) poor communication of information to other individuals, departments, or groups in time for it to be useful;

(ii) lack of understanding of where decisions are taken or goals set;

(iii) low involvement of other individuals, departments, or groups in reaching decisions;

(iv) lack of appreciation of the role of other individuals, departments, or groups in reaching goals or targets;

(v) failure to identify and use systems, methods, or techniques for specific activities;

(vi) absence of corrective action, following identification of weaknesses and problems;

(vii) lack of recognition of the role of training and follow-up.

The important aspect of methods used to counter these is the team building impact within each operational group. Each participant must have the opportunity to review their current roles, seek to change those aspects for which they perceive valid cause, and become more involved in those areas where they feel their inputs would have merit.

2 Developing performance indicators and objectives

Although the responsibilities should clarify what is to be performed, they do not define *how well* the tasks are expected to be performed. Performance indicators, therefore, are the means by which performance will be evaluated. To be meaningful they must be:

(a) **Measurable:** Indicators must lead to performance objectives which are quantifiable and tangible. Achievements in these areas must be recordable, verifiable, and observable. Quantity or quality of output, time schedules, costs, ratios, or percentages would be examples of measurable indicators.

(b) **Relevant:** Indicators must serve as a linkage between specific areas of responsibilities and the individual performance objectives to monitor achievement. They must describe what is the expected role of the position – the critical areas of performance.

(c) **Important:** Indicators need not be defined for every area of responsibility. They should be developed for those activities which have a significant impact on the results for the individual, department, and the organization.

The establishment of performance objectives provides clear direction and communication of expected levels of achievement. The process is a joint one – an interaction between the manager and his/her subordinates. If full commitment on the part of both parties is to be realized, the targets should be negotiated through a 'catch-ball' process, in the form of a performance 'contract'. Once the indicators have been agreed, the specific results desired need to be decided. The greater the participation, the greater the motivation to achieve. Agreed performance objectives should, therefore, contain the following ingredients:

(a)   Participatively developed.
(b)   Challenging but attainable.
(c)   Clear statements of performance expectations.
(d)   Within the individual's scope of control.

(a) **Participation** – An interaction, which leads to mutual agreement, provides a good exchange of ideas between the manager and his/her subordinates. The results are not a compromise but should be the outcome of a persuasive but logical presentation of why such an outcome is plausible. Discussions should be analytical – not emotional – and deal with both sides of an issue if there are significant differences. The crucial factors in examining the advantages of this approach are:

> Involvement → Commitment → Personal Responsibility → Higher drive to achieve

rather than:

> Imposition → Lack of acceptance → External responsibility → Lower drive to achieve

(b) **Challenge** – A well-set performance objective is one which is attainable but yet requires stretching. The achiever sets targets, which involve moderate risk. When the likelihood of success is 65 to 85 per cent, the inner sense of challenge is at its peak. As this probability decreases or increases from this range, the motive to achieve is reduced. The former makes the risk too great, since the target becomes perceived as unrealistic and self-esteem is lowered. The latter sets the risk as too low and if success is 'guaranteed', the pay-off value attached to attainment is reduced.

When individuals press for objectives that are either too low or too high, they tend to be motivated more by a fear of failure than the need to achieve. Those in this category either want the target to be fail-safe and, hence, be assured of success or else want to set a target so high that no one really takes their goals seriously.

To deal most effectively with either of these personalities, the performance objectives which are established should be of three levels: minimally acceptable, above average, and excellent. A person need not negotiate the minimal acceptable level since this is the least level of performance to maintain employment. The other levels can be discussed to arrive at realistic but challenging targets. Once they have been agreed upon, the choice of which path to follow is that of the subordinate – and the rewards can be similarly distributed.

(c) **Clarity of expectation** – The target should be objectively expressed and be tied to a specific time framework. Expressions such as 'approximate; minimum; maximum; adequate; none; as soon as possible' are vague and should be avoided. Descriptive, evaluative terms such as 'frequently; seldom; usually' etc. are also open to misinterpretation.

(d) **Scope of control** – The performance of the responsibility must be within the limits of authority that have been delegated. An individual cannot be reasonably held responsible for activities that cannot be directly controlled or influenced. For example, a production manager's performance objective of reviewing and accounting for the variation between budgeted and actual performance by the fifth working day of the month may not be adequately expressed, since the input for review may originate in data processing or accounting, rather than the manager's own department. If this is so, he may have no control over the budgeting data being available in time for a review on that date. A better indicator might be the time from receipt of the input to the submission of the analysis and recommendation.

3   Preparing action plans

It is clear that some form of action plan, perhaps in the form of a flow chart, bar chart or Gantt chart, is required to enable the objectives to be reached. The plans should stipulate action by the individuals concerned and be reviewed periodically against the milestones set down. For example:

- How will contributions made by individuals in team projects be evaluated?
- What action will be required to improve job performance?
- What are the criteria for promotion?
- What are the training needs to improve performance or prepare for promotion?
- What are the changes in the goals for the next performance period?

In order to effectively manage performance, an organization must have a performance management system for all its levels. The true translation of goals from the top to the bottom of the organization requires that one level's 'hows' become the next level's 'whats'. This interlocking or goal translation process should ensure the whole organization is working towards the same achievable mission, as was previously discussed in Chapter 3.

The need for teamwork

The complexity of most of the processes which are operated in industry, commerce and the services place them beyond the control of any one individual. The only efficient way to tackle process improvement or problems is through the use of some form of teamwork. The use of the team approach to improvement has many advantages over allowing individuals to work separately:

- a greater variety of complex issues may be tackled, which are beyond the capability of any one individual or even one department, by the pooling of expertise and resources,

- problems are exposed to a greater diversity of knowledge, skill, experience, and are solved more efficiently,

- the approach is more satisfying to team members and boosts morale and ownership through participation in decision making,

- improvement opportunities which cross departmental or functional boundaries can be addressed more easily, and the potential/actual conflicts are more likely to be identified and solved,

- the recommendations are more likely to be implemented than individual suggestions as the quality of decision making in **good teams** is high.

Most of these factors rely on the premise that people are most willing to support any effort in which they have taken part or helped to develop.

When properly managed and developed, teams improve processes, producing results quickly and economically. Teamwork throughout any organization is an essential component of total organizational excellence for it builds trust, improves communications and develops interdependence. Much of what has been taught previously in management has led to a culture in the West of independence, with little sharing of ideas and information. Knowledge is very much like organic manure – if it is spread around it will fertilize and encourage growth, if it is kept closed in, it will eventually fester and rot.

Teamwork devoted to process improvement changes the independence to interdependence through improved communications, trust and the free exchange of ideas, knowledge, data and information (Figure 10.1). The use of the face-to-face interaction method of communication, with a common goal, develops over time, the sense of dependence on each other. This forms a key part of any improvement process, and provides a methodology for employee recognition and involvement, through active encouragement in group activities.

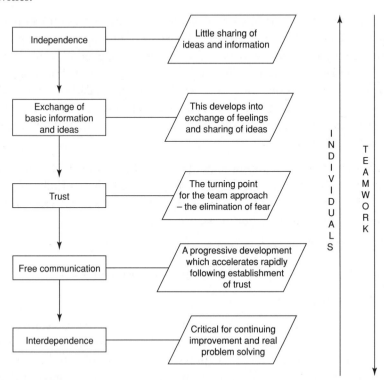

Figure 10.1 Independence to interdependence through teamwork

Teamwork provides an environment in which people can grow and use all the resources effectively and efficiently to make continuous improvements. As individuals grow, the organization grows. It is worth pointing out, however, that employees will not be motivated towards continual improvement in the absence of:

- commitment from top management,

- the right organizational 'climate',

- a mechanism for enabling individual contributions to be effective.

All these are focused essentially at enabling people to feel, accept, and discharge responsibility. More than one organization has made this part of their strategy – to 'empower people to act'. If one hears from employees comments such as, 'We know this is not the best way to do this job, but if that is the way management wants us to do it, that is the way we will do it', then it is clear that the expertise which exists at the point of operation has not been harnessed and the people do not feel responsible for the outcome of their actions. Responsibility and accountability foster pride, job satisfaction, and better work.

Empowerment to act is very easy to express conceptually, but it requires real effort and commitment on the part of all managers and supervisors to put into practice. Recognition that only partially successful but good ideas or attempts are to be applauded and not criticized is a good way to start. Encouragement of ideas and suggestions from the workforce, particularly through their involvement in team or group activities, requires investment. The rewards are total involvement, both inside the organization and outside through the supplier and customer chains.

Teamwork for process improvement has several components. It is driven by a strategy, needs a structure, and must be implemented thoughtfully and effectively. The strategy, which drives the improvement teams at the various levels, was outlined in Chapter 3, but in essence it comprises:

- the mission of the organization,

- the critical success factors, with key performance indicators,

- and the core processes.

The structure of having the top management team in a steering committee, the core processes being owned by process teams, which manage improvement projects requires some attention to be given to what makes people work well together, and what constitutes inspirational leadership.

Action centred leadership

During the 1960s John Adair was senior lecturer in military history and the Leadership Training Advisor at the Military Academy, Sandhurst, UK. Later, when assistant director of the Industrial Society, he developed what he called the action-centred leadership model, based on his experiences at Sandhurst where he had the responsibility to ensure that results in cadet training did not fall below a certain standard. He had observed that some instructors frequently achieved well above average results due to their own natural ability with groups and their enthusiasm. He developed this further into a team model which is the basis for the approach of the author and his colleagues to this subject.

In developing this model for teamwork and leadership, John Adair understood that for any group or team, big or small, to respond to leadership, they need a clearly defined *task,* and the response and achievement of that task is inter-related to the needs of the *team,* and the separate needs of the *individual members* of the *team* (Figure 10.2). The value of the overlapping circles is that it emphasizes the unity of leadership and the interdependence and multifunctional reaction to single decisions affecting any of the three areas.

Leadership tasks

Drawing upon the discipline of social psychology, John Adair developed and applied to training the functional view of leadership. The essence of this he distilled into the three inter-related but distinctive requirements of a leader. These are: to define and achieve the job or task, to build up and co-ordinate a team to do this, and to develop and satisfy the individuals within the team. The needs are shown in Figure 10.3.

Figure 10.2   Adair's model

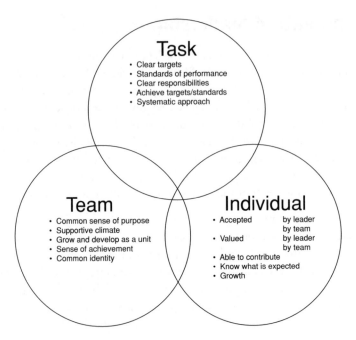

Figure 10.3 The leadership needs

- **Task needs** – The difference between a team and a random crowd is that a team has some common purpose, goal or objective, e.g. a football team. If a work team does not achieve the required results or meaningful results it will become frustrated. Organizations have a task: to make a profit, to provide a service, or even to survive. So anyone who manages others has to achieve results: production, marketing, selling or whatever. Achieving objectives is a major criterion of success.

- **Team needs** – To achieve these objectives the group needs to be held together. People need to be working in a co-ordinated fashion in the same direction. Teamwork will ensure that the team's contribution is greater than the sum of its parts. Conflict within the team must be used effectively, arguments can lead to ideas or to tension and lack of co-operation.

- **Individual needs** – Within working groups, individuals also have their own set of needs. They need to know what their responsibilities are, how they will be needed, how well they are performing. They need an opportunity to show their potential, take on responsibility and receive recognition for good work.

The task, team and individual functions for the leader are as follows:

Task functions – defining the task

- Making a plan
- Allocating work and resources
- Controlling quality and tempo of work
- Checking performance against the plan
- Adjusting the plan

Team functions – setting standards

- Maintaining discipline
- Building team spirit
- Encouraging, motivating, giving a sense of purpose
- Appointing sub-leaders
- Ensuring communication within the group
- Training the group

Individual functions – attending to personal problems

- Praising individuals
- Giving status
- Recognizing and using individual abilities
- Training the individual

The team leader's or facilitator's task is to concentrate on the small central area where all three circles overlap – the 'action to change' area. In this area the facilitator's or leader's task is to try to satisfy all three areas of need by achieving the task, building the team, and satisfying individual needs. If a leader concentrates on the task, e.g., in going all out for production schedules, while neglecting the training, encouragement and motivation of the team and individuals, (s)he may do very well in the short term. Eventually, however, the team members will give less effort than they are capable of. Similarly a leader who concentrates on only creating team spirit, while neglecting the task and the individuals, will not receive maximum contribution from the people. They may enjoy working in the team but they will lack the real sense of achievement which comes from accomplishing a task to the utmost of their collective ability.

So the leader/facilitator must try to achieve a balance by acting in all three areas of overlapping need. It is always wise to work out a list of required functions within the context of any given situation, based on a general agreement on the essentials. Here is Adair's original Sandhurst list on which may be based one's own information.

- **Planning**, e.g. seeking all available information.
  Defining group task, purpose or goal.
  Making a workable plan (in right decision-making framework).

- **Initiating**, e.g. briefing group on the aims and the plan.
  Explaining why aim or plan is necessary.
  Allocating tasks to group members.
  Setting group standards.

- **Controlling**, e.g. maintaining group standards.
  Influencing tempo.
  Ensuring all actions are taken towards objectives.
  Keeping discussion relevant.
  Prodding group to action/decision.

- **Supporting**, e.g. expressing acceptance of persons and their contribution.
  Encouraging group/individuals.
  Disciplining group/individuals.
  Creating team spirit.
  Relieving tension with humour.
  Reconciling disagreements or getting others to explore them.

- **Informing**, e.g. clarifying task and plan.
  Giving new information to the group, i.e. keeping them 'in the picture'.
  Receiving information from group.
  Summarizing suggestions and ideas coherently.

- **Evaluating,** e.g. checking feasibility of an idea.
  Testing the consequences of a proposed solution.
  Evaluating group performance.
  Helping the group to evaluate its own performance against standards.

A checklist is given in Table 10.1 which should assist the team leader to measure the progress against the required functions of fulfilling the task, maintaining the team and growing the people.

Team processes

The team process is like any other process, it has inputs and outputs. High performing teams have three main attributes: high task fulfilment, high team maintenance, and low self-orientation. These may be subdivided as follows:

Task fulfilment

Initiating	Ideas, solutions, defining problems, suggesting procedures, proposing tasks or goals.
Information seeking	Facts, opinions, suggestions, ideas.
Information giving	Facts, opinions, suggestions, ideas.
Clarifying	Analysing implications of information or ideas, interpreting, defining terms, indicating alternatives.
Summarizing	Reviewing, drawing together ideas/information.
Testing for consensus	Checking readiness for decision, checking agreements.

Team maintenance

Encouraging	Supporting contributions, stimulating contributions, being friendly and responsive.

Table 10.1   Task–team–individual checklist

Task
1.   Are the targets clearly set out?
2.   Are there clear standards of performance?
3.   Are available resources defined?
4.   Are responsibilities clear?
5.   Are resources fully utilized?
6.   Are targets/standards being defined?
7.   Is a systematic approach being used?

Team
1.   Is there a common sense of purpose?
2.   Is there a supportive climate?
3.   Is the unit growing and developing?
4.   Is there a sense of corporate achievement?
5.   Is there a common identity?
6.   Does the team know and respond to the leader's vision?

Individual
1.   Is each individual accepted by the leader/team?
2.   Is each individual involved by leader/team?
3.   Is each individual able to contribute?
4.   Does each individual know what is expected in relation to the task and by the team?
5.   Does each individual feel a part of the team?
6.   Does each individual feel valued by the team?
7.   Is there evidence of individual growth?

Setting standards	Suggesting standards for group working, reviewing against these, evaluating group success.
Expressing group feelings	Observing, understanding and expressing group emotions, reducing tension, mediating, recognizing conflicts and encouraging exploration of differences.
Compromising	Giving weight to others' views, commitment to best solution.
Gatekeeping	Keeping communications open, facilitating participation, suggesting procedures for sharing discussion.

Self orientation

Blocking	Interposing a difficulty without alternative or reasoning.
Aggressiveness	Attacking, over painting the picture to stir up feelings, exaggerating.
Dominating	Asserting authority or superiority in manipulating group, refusing to budge.
Forming cliques	Forming sub-groups for protection or support.
Special pleading	Speaking for special interests as a cover for personal interest.
Seeking sympathy	Drawing attention, attempting to gain sympathy.
Withdrawing	Opting out or getting behind stronger members.
Wasting time	Various diversions for self orientated reasons.
Not listening	Ignoring suggestions or closing off hearing when others are speaking.

These may be used in various ways to construct a 'team behaviour checklist', which may be used by team facilitators or observers to rate the team performance. A second review document (see Table 10.2) may be used by individuals to rate the various aspects of a team meeting.

Stages of team development

Original work by Tuckman suggested that when teams are put together, there are four main stages of team development, the so-called forming (awareness), storming (conflict), norming (co-operation), and performing (productivity). The characteristics of each stage and some key aspects to look out for in the early stages are given below:

Table 10.2   Team meeting review

	10	5	0	
1. Goal clear and agreed				Goal unclear
2. Previous agreements complete				Partially or not at all
3. We listened to each other				No awareness of listening
4. Right people present				Team not correctly composed
5. Leadership needs creatively met				Drifting or dominating
6. Open & trusting atmosphere				Distrust & defensiveness
7. Time used efficiently				Time wasted
8. Systematic tools used				Lack of systematic approach
9. Agreements reached (what/who/when) & documented				Verbal agreements or none
10. Consensus decisions				Authoritarian or other
11. I was able to express my opinion				No opportunity
12. Opinions could be questioned				Opinions untouchable
13. Opinions distinguished from facts				Mixed & not aware of it
14. Everyone involved				Some not involved
15. Challenging, rewarding, committed atmosphere				Flat & lifeless

Unacceptable   0
Must be improved   1, 2, 3
Fair   4, 5, 6
Good   7, 8, 9
Excellent   10

Forming – awareness

Characteristics:

- feelings, weaknesses and mistakes are covered up,

- people conform to established lines,

- little care is shown for others' values and views,

- there is no shared understanding of what needs to be done.

Watch out for:

- increasing bureaucracy and paperwork,

- people confining themselves to defined jobs,

- the 'boss' is ruling with a firm hand.

Storming – conflict

Characteristics:

- more risky, personal issues are opened up,

- the team becomes more inward looking,

- there is more concern for the values, views and problems of others in the team.

Watch out for:

- the team becomes more open, but lacks the capacity to act in a unified, economic and effective way.

Norming – co-operation

Characteristics:

- confidence and trust to look at how the team is operating,

- a more systematic and open approach, leading to a clearer and more methodical way of working,

- greater valuing of people for their differences,

- clarification of purpose and establishing of objectives,

- systematic collection of information,

- considering all options,

- preparing detailed plans,

- reviewing progress to make improvements.

Performing – productivity

Characteristics:

- flexibility,

- leadership decided by situations, not protocol,

- everyone's energies utilized,

- basic principles and social aspects of the organization's decisions considered.

The team stages, the task outcomes, and the relationship outcomes are shown together in Figure 10.4. This model, which has been modified from Kormanski and Mozenter, may be used as a framework for the assessment of team performance. The issues to look for are:

- How is leadership exercised in the team?

- How is decision making accomplished?

- How are team resources utilized?

- How are new members integrated into the team?

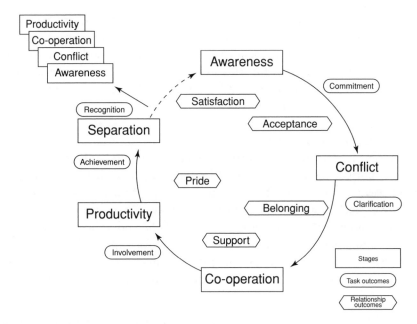

Figure 10.4   Team stages and outcomes.
Source: modified from Kormanski, C. and Mozenter, A. (1987) A New Model of Team Building, a technology for today and tomorrow, *The 1987 Annual Developing Human Resources Conference*, University Association of San Diego

Teams, which go through these stages successfully, should become effective teams and display:

Clear objectives and agreed performance goals

No group of people can be effective unless they know what they want to achieve, but it is more than knowing what the objectives are. People are only likely to be committed to them if they can identify with and have ownership of them – in other words, objectives and performance goals are agreed by team members. Often this agreement is difficult to achieve but experience shows that it is an essential prerequisite for the effective group.

Openness and confrontation

If a team is to be effective then the members of it need to be able to state their views, their differences of opinion, interests and problems, without fear of ridicule or retaliation. No teams work effectively if there is a cut-throat atmosphere, where members become less willing or able to express themselves openly – much energy, effort and creativity is lost.

Support and trust

Support naturally implies trust among team members. Where individual group members do not feel they have to protect their territory or job, and feel able to talk straight to other members – about both 'nice' and 'nasty' things – then the opportunity exists for trust to be shown. Based on this trust people can talk freely about their fears and problems and receive from others help which they need to be more effective.

Co-operation and conflict

When there is an atmosphere of trust, members are more ready to be involved and committed. Information is shared rather than hidden. Individuals listen to the ideas of others and build on them. People find ways of being more helpful to each other and the group generally. Co-operation causes high morale – individuals accept each other's strengths and weaknesses and contribute from their pool of knowledge and skill. All abilities, knowledge and experience are fully utilized by the group; individuals have no inhibitions about using other people's abilities to help solve their problems, which are shared.

Allied to this, conflicts are seen as a necessary and useful part of organizational life. The effective team works through issues of conflict and uses the results to help achieve objectives. Conflict prevents teams from becoming complacent and lazy and often they generate new ideas.

Good decision making

As mentioned earlier, objectives need to be clearly and completely understood by all members before good decision making can commence. In making decisions effective, teams develop the ability to collect information quickly then discuss the alternatives openly. They become committed to their decisions and ensure action quickly.

Appropriate leadership

Effective teams have a leader whose responsibility it is to achieve results through the efforts of a number of people. Power and authority can be applied in many ways and team members often differ on the style of leadership they prefer. Collectively, teams may come to different views of leadership but, whatever their view, the effective team usually sorts through the alternatives in an open and honest way.

Review of the team process

Effective teams understand, not only the group's character and its role in the organization, but how it makes decisions, deals with conflicts, etc. The team process allows the team to learn from experiences and to consciously improve teamwork. There are numerous ways of looking at team processes – use of an observer, by a team member giving feedback or by the whole group discussing their performance.

Sound inter-group relationships

No human being or group is an island, they need the help of others. An organization will not achieve maximum benefit from a collection of improvement teams which are effective within themselves, but which fight amongst each other.

Individual development opportunities

High performance teams seek to pool the skills of individuals and it necessarily follows that they pay attention to development of individual skills and try to provide opportunities for individuals to grow and learn and, of course, have *fun*.

These ideas are not particularly new but are very applicable and useful in the management of teams for improvement, just as Newton's theories on gravity still apply!

Personality types and the MBTI

No one person has a monopoly of 'good' characteristics. Attempts to list the qualities of the ideal manager, for example, demonstrate why that paragon cannot exist. This is because many of the qualities are mutually exclusive, for example:

Highly intelligent	v	Not *too* clever
Forceful and driving	v	Sensitive to people's feelings
Dynamic	v	Patient
Fluent communicator	v	Good listener
Decisive	v	Reflective

Although no individual can encompass all these and other desirable qualities, a team often does.

A powerful aid to team development is the use of the Myers Briggs Type Indicator (MBTI). This is based on an individual's preferences on four scales for:

- giving and receiving 'energy',

- gathering information,

- making decisions,

- handling the outer world.

Its aim is to help individuals understand and value themselves and others, in terms of their differences as well as their similarities. It is well researched and non-threatening when used appropriately.

The four MBTI preference scales, which are based on Jung's theories of psychological types, represent two opposite preferences:

- **Extroversion–Introversion:** how we prefer to give/receive energy or focus our attention.

- **Sensing–iNtuition:** how we prefer to gather information.

- **Thinking–Feeling:** how we prefer to make decisions.

- **Judgement–Perception:** how we prefer to handle the outer world.

To understand what is meant by preferences, the analogy of left and right handedness is useful. Most people have a preference to write with either their left or their right hand. When using the preferred hand, they tend not to think about it, it is done naturally. When writing with the other hand, however, it takes longer, needs careful concentration, seems more difficult, but with practice would no doubt become easier. Most people **can** write with and use both hands, but tend to prefer one over the other. This is similar to MBTI psychological preferences; most people are able to use both preferences at different times, but will indicate a preference on each of the scales.

In all, there are eight possible preferences – E or I, S or N, T or F, J or P, i.e. two opposites for each of the four scales. An individual's **type** is the combination of interaction of the four preferences. It can be assessed initially by completion of a simple questionnaire. Hence if each preference is represented by its letter, a person's type may be shown by a four letter code, there are sixteen in all. For example, ESTJ represents an **extrovert** (E) who prefers to gather information with **sensing** (S), prefers to make decisions using **thinking** (T) and who prefers a **judging** (J) attitude towards the world, i.e. prefers to make decisions rather than continue to collect information. The person with opposite preferences on all four scales would be an INFP, an introvert, who prefers intuition for perceiving, feelings, or values for making decisions, and who prefers to maintain a perceiving attitude towards the outer world.

The questionnaire, its analysis and feedback must be administered by a qualified MBTI practitioner, who may also act as external facilitator to the team in its forming and storming stages.

Type and teamwork

With regard to teamwork, the preference types and their interpretation are extremely powerful. The **extrovert** prefers action and the outer world, whilst the **introvert** prefers ideas and the inner world.

Sensing–thinking types are interested in facts, analyse facts impersonally, and use a step-by-step process from cause to effect, premise to conclusion. The **sensing–feeling** combination, however, are interested in facts, analyse facts personally, and are concerned about how things matter to themselves and others.

Intuition–thinking types are interested in possibilities, analyse possibilities impersonally, and have theoretical, technical, or executive abilities. On the other hand, the **intuition–feeling** combination is interested in possibilities, analyses possibilities personally, and prefers new projects, new truths, things not yet apparent.

Judging types are decisive and planful, they live in orderly fashion, and like to regulate and control. **Perceivers** on the other hand are flexible, live spontaneously, and understand and adapt readily.

As we have seen, an individual's type is the combination of four preferences on each of the scales. There are sixteen possible combinations of the preference scales and these may be displayed on a **type table** (Figure 10.5). If the individ-

ISTJ	ISFJ	INFJ	INTJ
ISTP	ISFP	INFP	INTP
ESTP	ESFP	ENFP	ENTP
ESTJ	ESFJ	ENFJ	ENTJ

Figure 10.5 MBTI type table form
Source: Isabel Briggs Myers (1987) *Introduction to Type*

uals within a team are prepared to share with each other their MBTI preferences, this can dramatically increase understanding and frequently is of great assistance in team development and good team working. The similarities and differences in behaviour and personality can be identified. The assistance of a qualified MBTI practitioner is absolutely essential in the initial stages of this work.

Process improvement and type preferences

The MBTI preferences may be used by an individual or a team in a step-by-step process for problem solving or improvement. The process improvement model represented in Figure 10.6 is straightforward but can be difficult to use because people tend to skip over those steps which require them to use their non-preferences. For example, information tends to be gathered by the preferred function (S or N) and decisions made by the preferred function (T or F). So a strongly ST type will spend much time gathering facts (S) and thinking logically through the decision process (T) with perhaps insufficient attention being given to other possibilities (N) and the impact on people (F). If the size of each letter represents a unit of time, the ST's problem solving method may be represented in Figure 10.7, in which the Z pattern of the model is not

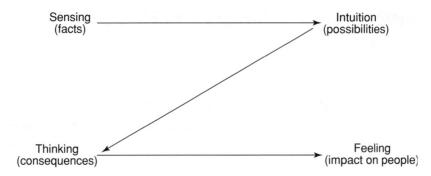

Figure 10.6 Team task analysis: process improvement model

followed. Problems, solutions, and decisions are likely to be improved if all the preferences are used. Until individuals master the process of spending time in their non-preferred functions (i.e. type development), it may be wise to consult others of opposite preferences when tackling important problems or making vital decisions.

Clearly, this has great implications for teamwork and requires that team members share their MBTI preferences or types.

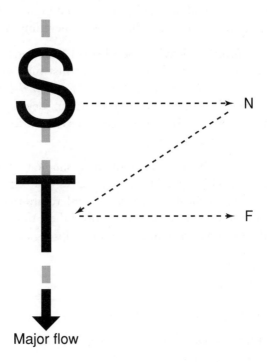

Figure 10.7 Relative time spent on each aspect of problem-solving model by ST type

Interpersonal relations – FIRO-B and the Elements

The FIRO-B (Fundamental Interpersonal Relations Orientation-Behaviour) is a powerful psychological instrument, which can be used to give valuable insights into the needs individuals bring to their relationships with other people. The instrument assesses needs for **inclusion, control** and **openness** and thereby offers a framework for understanding the dynamics of interpersonal relationships.

Use of the FIRO instrument helps individuals to be more aware of how they relate to others and to become more flexible in this behaviour. Consequently it enables people to build more productive teams through better working relationships.

Since its creation by William Schutz in the 1950s, to predict how military personnel would work together in groups, the FIRO-B instrument has been used throughout the world by managers and professionals to look at management and decision making styles. Through its ability to predict areas of probable tension and compatibility between individuals, the FIRO-B is a highly effective team building tool which can aid in the creation of the positive environment in which people thrive and achieve improvements in performance.

The theory underlying the FIRO-B incorporates ideas from the work of Adomo, Fromm and Bion and it was first fully described in Schutz's book, *FIRO: A Three Dimensional Theory of Personal Behaviour* (1958). In his more recent book *The Human Element*, Schutz developed the instrument into a series of 'elements', B, F, S, etc. and offers strategies for heightening our awareness of ourselves and others.

The FIRO-B takes the form of a simple-to-complete questionnaire the analysis of which provides scores that estimate the levels of behaviour with which the individual is comfortable, with regard to his/her needs for inclusion, control and openness. Schutz described these three dimensions in the form of the decisions we make in our relationships regarding whether we want to be:

- 'in' or 'out' – inclusion
- 'up' or 'down' – control
- 'close' or 'distant' – openness

The FIRO-B estimates our unique level of needs for each of these dimensions of interpersonal interaction.

The instrument further divides each of these dimensions into:

Table 10.3   The FIRO-B interpersonal dimensions and aspects

	Inclusion	Control	Openness
Expressed behaviour	Expressed inclusion	Expressed control	Expressed openness
Wanted behaviour	Wanted inclusion	Wanted control	Wanted openness

Modified from: W. Schutz (1978) *FIRO Awareness Scales Manual*, Palo Alto, CA, Consulting Psychologists Press.

i)   the behaviour we feel most comfortable **exhibiting towards** other people – **expressed** behaviours, and

ii) the behaviour we **want from** others – **wanted** behaviours.

Hence, the FIRO-B 'measures', on a scale of 0–9, each of the three interpersonal dimensions in two aspects (Table 10.3).

The **expressed** aspect of each dimension indicates the level of behaviour the individual is most comfortable with towards others, so high scores for the expressed dimensions would be associated with:

High scored expressed behaviours

Inclusion   Makes efforts to include other people in his/her activities – tries to belong to or join groups and to be with people as much as possible.

Control   Tries to exert control and influence over people and tell them what to do.

Openness   Makes efforts to become close to people – expresses friendly open feelings, tries to be personal and even intimate.

Low scores would be associated with the opposite expressed behaviour.

The **wanted** aspect of each dimension indicates the behaviour the individual prefers others to adopt towards him/her, so high scores for the wanted dimensions would be associated with:

High scored wanted behaviours

Inclusion   Wants other people to include him/her in their activities – to be invited to belong to or join groups (even if no effort is made by the individual to be included).

Control   Wants others to control and influence him/her and be told what to do.

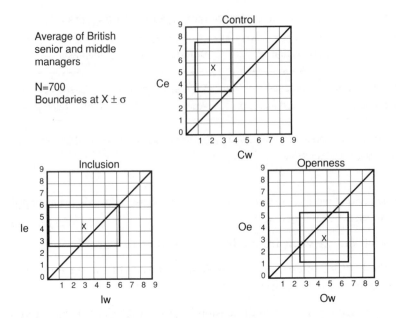

Figure 10.8 Typical manager profiles (FIRO-B)

Openness Wants others to become close to him/her and express friendly, open, even affectionate feelings.

Low scores would be associated with the opposite wanted behaviours.

It is interesting to look at typical manager FIRO-B profiles, based on their scores for the six dimensions/aspects in Table 10.3. Figure 10.8 shows the average of a sample of 700 middle and senior managers in the UK with boundaries at one sigma, plotted on expressed/wanted scales for the three dimensions.

On average the managers show a higher level of expressed inclusion – including people in his/her activities – than wanted inclusion. Similarly, and not surprisingly perhaps, expressed control – trying to exert influence and control over others – is higher in managers than wanted control. When it comes to openness, the managers tend to want others to be open, rather than be open themselves.

It is even more interesting to contemplate these results when one considers the demands of some of the recent popular management programmes, such as total quality management, employee involvement, and self-directed teams. These tend to require from managers certain behaviours, for example, lower levels of expressed control and higher levels of wanted control, so that the people feel empowered. Similarly, managers are encouraged to be more open. These, however, are **opposite** to the apparent behaviours of the sample of managers shown graphically in Figure 10.8. It is not surprising then that TQM failed in some organizations where managers were being asked to empower employees and be more open – and who can argue against that – yet their basic underlying needs caused them to behave in the opposite way.

Understanding what drives these behaviours is outside the scope of this book but other FIRO and Element instruments can help individuals to further develop understanding of themselves and others. FIRO and Schutz's Elements, instruments for measuring **feelings** (F) **self-concept** (S), can deepen the awareness of what lies behind our behaviours with respect to inclusion, control and openness. The reader is advised to undertake further reading and seek guidance from a trained administrator of these instruments, but the overall relationship between the B and F instruments is given below:

Behaviours related to:	Feelings related to:
Inclusion	Significance
Control	Competence
Openness	Likeability

Issues around control behaviour then may arise because of underlying feelings about competence. Similarly, underlying feelings concerning significance may lead to certain inclusion behaviours.

FIRO-B in the work place

The inclusion, control and openness dimensions form a cycle (Figure 10.9) which can help groups of people to understand how their individual and joint behaviour develops as teams are formed. Given in Table 10.4 are the considerations, questions and outcomes under each dimension. If inclusion issues are resolved first it is possible to progress to dealing with the control issues, which in turn must be resolved if the openness issues are to be dealt with successfully. As a team develops, it travels around the inclusion, control and openness cycle time and time again. If the issues are not resolved in each dimension, further progress in the next dimension will be hindered – it is difficult to deal with issues of control if unresolved inclusion issues are still around and people do not know whether they are 'in' or 'out' of the group. Similarly it is difficult to be open if it is not clear where the power base in the group lies.

This I–C–O cycle has led to the development by the author and his colleagues of an 'openness model' which is in three parts. Part 1 is based on the premise that to participate productively in a team individuals must firstly be involved and then committed. Figure 10.10 shows some of the questions

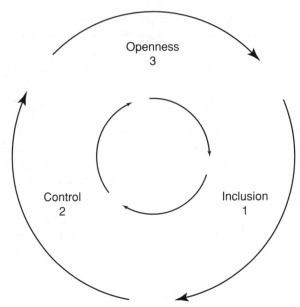

Figure 10.9   The inclusion, control and openness cycle

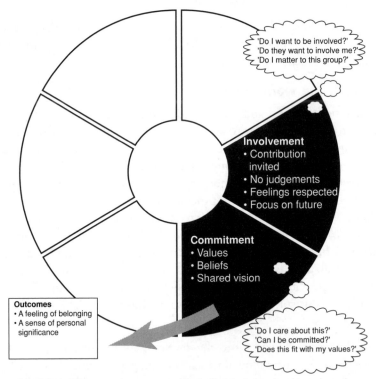

Figure 10.10   The openness model, Part 1 Inclusion: involvement–inviting contribution–responding

Table 10.4 Considerations, questions and outcomes for the FIRO-B dimensions

Dimension	Considerations	Some typical questions	If resolved we get:	If not resolved we get:
Inclusion	Involvement – how much you want to include other people in your life and how much attention and recognition you want.	Do I care about this? Do I want to be involved? Does this fit with my values? Do I matter to this group? Can I be committed? ...leading to... Am I 'in' or 'out'?	A feeling of belonging A sense of being recognized and valued Willingness to become committed	A feeling of alienation A sense of personal insignificance No desire for commitment or involvement
Control	Authority, responsibility, decision making, influence.	Who is in charge here? Do I have power to make decisions? What is the plan? When do we start? What support do I have? What resources do I have? ...leading to... Am I 'up' or 'down'?	Confidence in self and others Comfort with level of responsibility Willingness to belong	Lack of confidence in leadership Discomfort with level of responsibility – fear of too much – frustration with too little 'Griping' between individuals
Openness	How much we are prepared to express our true thoughts and feelings with other individuals.	Does she like me? Will my work be recognized? Is he being honest with me? How should I show appreciation? Do I appear aloof? ...leading to... Am I 'open' or 'closed'	Lively and relaxed atmosphere Good-humoured interactions Open and trusting relationships	Tense and suspicious atmosphere Flippant or malicious humour Individuals isolated

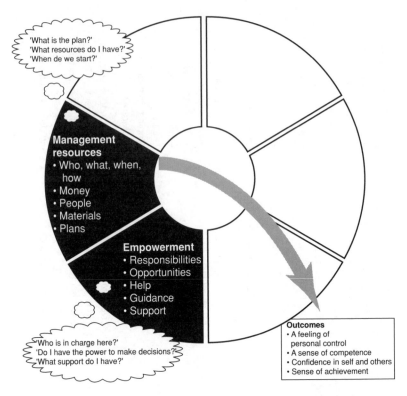

Figure 10.11   The openness model, Part 2 Control: choice–influence–power

which need to be answered and the outcomes from this stage. Part 2 deals with the control aspects of empowerment and management and Figure 10.11 summarizes the questions and outcomes. Finally Part 3, summarized in Figure 10.12, ensures openness through acknowledgement and trust. The full openness cycle (Figure 10.13) operates in a clockwise direction so that trust leads to more involvement, further commitment, increased empowerment, etc. Of course, if progress is not made round the cycle and trust is replaced by fear, it is possible to send the whole process into reverse – a negative cycle of suspicion fault-finding, abdication and confusion (Figure 10.14). Unfortunately this will be recognized as the culture in some organizations where the focus of enquiry is 'what has gone wrong?' leading to 'whose fault was it?'

Fortunately, organizations and individuals seem keen to learn ways to change these negative communications that sour relationships, dampen personal satisfaction and reduce productivity. The inclusion–control–openness cycle is a useful framework for helping teams to pass successfully through the

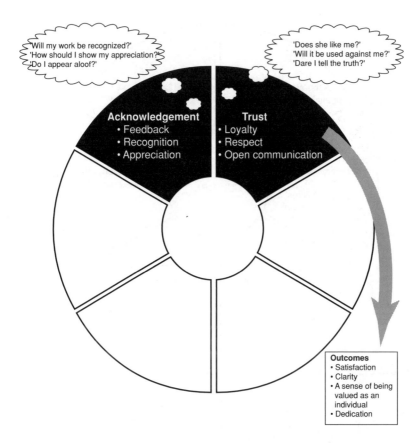

Figure 10.12 The openness model, Part 3 Openness: expression of true thoughts and feelings with respect for self and others

forming and storming stages of team development. As teams are disbanded for whatever reason, the process reverses and the first thing which goes is the openness.

The five 'A' stages for teamwork

The awareness provided by the use of the MBTI and FIRO-B instruments helps people to appreciate their own uniqueness and the uniqueness of others – the foundation of mutual respect and for building positive, productive and high performing teams.

For any of these models or theories to benefit a team, however, the individuals within it need to become **aware** of the theory, e.g., the MBTI or FIRO-B. They then need to **accept** the principles as valid, **adopt** them for

Figure 10.13   The openness model

Figure 10.14   The negative cycle

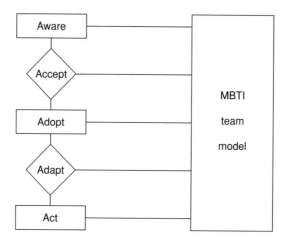

Figure 10.15   The five 'A' stages for teamwork

themselves in order to **adapt** their behaviour accordingly. This will lead to individual and team **action** (Figure 10.15).

In the early stages of team development particularly, the assistance of a skilled facilitator to aid progress through these stages is necessary. This is often neglected, causing failure in so many team initiatives from which the net output turns out to be lots of nice warm feelings about 'how good that team workshop was a year ago', but the nagging reality that no action came out and nothing has really changed.

Many teams fail because it is assumed that after initial training they are fully able to work effectively and deliver rapid results. This is rarely the case; they are still very much 'learner-drivers' trying to apply a new way of working. At this critical stage, it is vital that they have easy access to help. A small group of facilitators should be trained to fulfil this role in order to:

- provide a resource to improvement teams to assist them to work effectively

- improve the ability of people to work in teams

- help to break down inter-departmental barriers

- ensure that a systematic approach is used

- help team leaders to develop their leadership abilities

- ensure that maximum measured benefit is derived from the investment made.

Steering committees and teams

The achievement of total organizational excellence takes considerable time and ability. It must be given the status of a senior executive project because of the need for full integration with the organization's strategy, operating philosophy and management systems. It may require an extensive review and maybe substantial revision of existing systems of management and ways of operating. Fundamental questions may have to be asked, such as: Do the managers have the necessary authority, capability, and time to carry this through?

Any review of existing management and operating systems will inevitably open many cans of worms and uncover problems that have been successfully buried and smoothed over – perhaps for years. Authority must be given to those charged with following through with actions that they consider necessary to achieve the goals. The commitment will be continually questioned and will be weakened, perhaps destroyed by the failure to delegate authoritatively.

The following steps are suggested in general terms. Clearly, different types of organizations will need to make adjustments to the detail, but the component parts are the basic requirements.

A disciplined and systematic approach to organizational excellence may be established in an executive steering committee which should meet at least monthly to review strategy, implementation progress and improvement. It should be chaired by the chief executive and the members should include the top management team and the chairs of any 'site' steering committees, or core process teams, depending on the size of the organization. The objectives of the committee are to:

- provide overall strategic direction for the organization;

- establish plans in each ' site' or division;

- set up and review the teams which will own the key or core business processes;

- review and revise plans for implementation

The core process teams and any site steering committees should also meet monthly, shortly before the committee meetings. Every senior manager should be a member of at least one core process team. This system provides the 'top-down' support for employee involvement in process management and development, through continuous improvement. It also ensures that the commitment at the top is communicated effectively through the organization.

The three-tier approach of steering committee, process teams, and continuous improvement teams allows the former to concentrate on strategy, rather than become a senior problem-solving group. Progress is assured if the process team chairperson is required to present a status report at each meeting.

The steering committee and process teams will control the improvement teams and have the responsibility for:

- the selection of projects for improvement teams;

- providing an outline and scope for each project to give to the teams;

- the appointment of team members and leaders;

- monitoring and reviewing the progress and results from each project.

As the focus of this work will be the selection of projects, some attention will need to be given to the sources of nominations. Projects may be suggested by:

i) steering committee members, representing their own departments, process teams, their suppliers or their customers, internal and external;

ii) continuous improvement teams;

iii) quality circles or similar (if in existence);

iv) suppliers;

v) customers.

The process team members must be given the responsibility and authority to represent their part of the organization in the process. The members must also feel that they represent the team to the rest of the organization. In this way the team will gain knowledge and respect and be seen to have the authority to act in the best interests of the organization, with respect to their process.

The actual running of improvement teams involves several factors:

- team selection and leadership

- team objectives

- team meetings

- team assignments

- team dynamics

- team results and reviews.

Team selection and leadership

The most important element of a team is its members. People with knowledge and experience relevant to the process and/or solving problems are clearly required, but there should be a limit of five to ten members to keep the team small enough to be manageable, but allow a good exchange of ideas. Membership should include appropriate people from groups outside the operational and technical areas directly 'responsible' for the problem, if their involvement is relevant or essential. In the selection of team members, it is often useful to start with just one or two people concerned directly with the problem. If they try to draw flow charts (see Chapter 5) of the processes involved, the requirement to include other people, in order to understand the process and complete the charts, will aid the team selection. This method will also ensure that all those who can make a significant contribution to the improvement process are represented.

The team leader has a primary responsibility for team management and maintenance and his/her selection and training is crucial to success. The leader need not be the highest ranking person in the team, but must be concerned about accomplishing the team objectives (this is sometimes described as 'task concern') and the needs of the members (often termed 'people concern'). Weakness in either of these areas will lessen the effectiveness of the team in making improvements. Team leadership training should be directed at correcting deficiencies in these crucial aspects.

Team objectives

At the beginning of any project and at the start of every meeting, the objectives should be stated as clearly as possible by the leader. This can take the simple form, 'This meeting is to continue the discussions from last Tuesday on the provision of current price data from salesmen to invoice preparation, and to generate suggestions for improvement in its quality'. Project and/or meeting objectives enable the team members to focus thoughts and efforts on the aims, which may need to be restated if the team becomes distracted by other issues.

Team meetings

An agenda should be prepared by the leader and distributed to each team member before every meeting. This should include the following information:

- meeting place, time and how long it will be;
- a list of members (and co-opted members) expected to attend;
- any preparatory assignments for individual members or groups;
- any supporting material to be discussed at the meeting.

A suggested process for team discussions is given in Figure 10.16.

Early in a project the leaders should orient the team members in terms of the approach, methods, and techniques they will use to solve the problem. This may require a review of the:

- systematic approach (Chapter 9);

- procedures and rules for using some of the basic tools, e.g. brainstorming – no judgement of initial ideas;

- role of the team in the overall continuous improvement process;

- authority of the team.

A team secretary should be appointed to take the minutes of meetings and distribute them to members, as soon as possible after each meeting. The minutes should not be formal, but reflect decisions and carry a clear statement of the action plans together with assignments of tasks. They may be hand-written initially, copied and given to team members at the end of the meeting, to be followed later by a more formal document which will be seen by any member

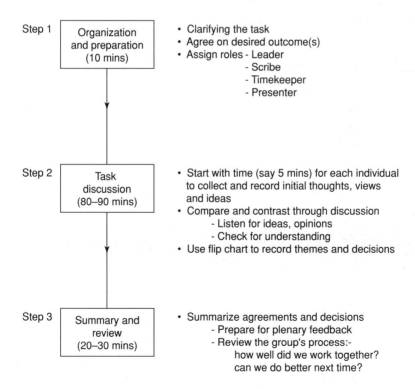

Figure 10.16 Suggested process for team discussions

of staff interested in knowing the outcome of the meeting. In this way the minutes form an important part of the communication system involving other teams or people involved in some way with the process.

Team assignments

It is never possible to solve problems by meetings alone. What must come out of those meetings is a series of action plans which assign specific tasks to team members. This is the responsibility of the team leader. Agreement must be reached regarding the responsibilities for individual assignments, together with the time scale, and this must be made clear in the minutes. Task assignments must be decided while the team is together and not by separate individuals in after-meeting discussions.

Team dynamics

In any team activity the interactions between the members is vital to success. If solutions to problems are to be found, the meetings and ensuing assignments should assist and harness the creative thinking process. This is easier said than done because many people have either not learned or been encouraged to be innovative. The team leader clearly has a role here to:

- create a 'climate' for creativity;
- encourage all team members to speak out and contribute their own ideas or build on others;
- allow differing points of view and ideas to emerge;
- remove barriers to idea generation, e.g. incorrect preconceptions which are usually destroyed by asking 'why?';
- support all team members in their attempts to become creative.

In addition to the team leader's responsibilities, the members have roles to:

- prepare themselves well for meetings, by collecting appropriate data or information (**facts**) pertaining to a particular problem;
- share ideas and opinions;
- encourage other points of view;
- listen 'openly' for alternative approaches to a problem or issue;
- help the team determine the best solutions;
- reserve judgement until all the arguments have been heard *and* fully understood;

- accept individual responsibility for assignments and group responsibility for the efforts of the team.

Team results and reviews

Teams function most effectively when the results of their projects are communicated and acted upon. Regular feedback to the teams, via their leaders, will assist them to focus on project objectives and review progress.

Reviews also help to deal with certain problems which may arise in teamwork. For example, certain members may be concerned more with their own personal objectives than those of the team. This may result in some manipulation of the team process to achieve different goals, resulting in the team splitting apart through self-interests. If recognized, the review can correct this effect and demand greater openness and honesty.

A different type of problem is the failure of certain members to contribute evenly and take their share of individual and group responsibility. Allowing other people to do their work results in an uneven distribution of effort, and bitterness. The review should make sure that all members have assigned and specific tasks, and perhaps lead to the documentation of duties in the minutes. A team roster may even help.

A third area of difficulty, which may be improved by reviewing progress, is the ready-fire-aim syndrome of action before analysis. This often results from team leaders being too anxious to deal with a problem. A review should allow problems to be redefined adequately and expose the real cause(s). This will release the trap the team may be in of doing something before they really know what should be done. The review will provide the opportunity to rehearse the steps in the systematic approach:

- **Record data:** all processes can and should be measured and all measurements should be recorded.

- **Use data:** if data is recorded and not used it will be abused.

- **Analyse data systematically:** data analysis should be carried out using the basic tools (Chapter 9).

- **Act on the results:** recording and analysis of data without action leads to frustration.

Team performance and culture

The failure to address the culture of the organization is frequently the reason for many management initiatives either having limited success or failing altogether. Understanding the culture of the organization and using that under-standing to successfully map the transitions needed to accomplish successful change will be an important part of the journey towards excellence.

There is widespread recognition today that major change initiatives will not succeed without a culture of good teamworking and co-operation at all levels in organizations. Much training effort has therefore been expended in trying to establish good teamwork principles and practice in the workforce. These efforts generally meet with some success in the short-term but in the longer-term the effects often do not persist. There can be a number of reasons for this:

- The senior management does not embrace the principles of teamworking for themselves.

- Major attention continues to be focused on the task, with relatively little focus on the people who are to carry it out.

- Functional management structures do not lend themselves readily to cross-functional co-operation.

- The principles are not practised at all levels in the organization, leading eventually to cynicism and low morale at the 'coal-face'.

Breaking down the departmental barriers is essential to the spread of team-working within the organization. The experience of the author and his colleagues shows that this will not be successful unless it is addressed by the senior management, starting with themselves. Thus a management team that demonstrates good teamworking itself and widely supports and encourages it in the rest of the workforce is likely to be more successful in bringing about change. In doing this it will generate a culture of higher co-operation and productivity throughout the organization.

Much has been written on the subject of team behaviour and performance – notably by some of the authors already mentioned in this chapter. All of the various approaches have value in shedding light on different aspects of the way individuals are likely to function as a group. They recognize that people are diverse and that diversity, not uniformity, is essential for the achievement of team goals. So most approaches tend to focus on managing, or 'balancing' the diversity in the group as a whole: Adair, for instance, advocates attending to, and balancing, the 'needs' of the task, team and individual.

The approach presented in this chapter recognizes that the first requirement is to assemble those with the ability to carry out the desired tasks. Balancing different profiles in a team is of little value if the individuals do not possess the right skills and talents for the job. An important next stage is the recognition that the performance of any team is the sum of all the paired interactions within the team. A team does not fail because of its diversity but because of failure of pairs of individuals to function properly together. These failures are caused by rigidities between individuals that do not get resolved, sometimes even after very long periods of time. For the whole team to be successful, *every* pair interaction must work effectively. Not until this has been achieved can the other aspects of team performance follow, namely group identity and efficient task achievement. This is where the MBTI and FIRO-B instruments can be so helpful to improving team performance and achievement.

The success of any strategic change initiative will depend critically on the successful management of the factors set out in general terms above. To achieve these, attention will need to be paid to the following:

- The role and composition of the team leading the change.

- The energy, commitment and vision of the individuals forming the leadership team.

- The inter-relationships of the team members with one another.

- The ability of the leaders to mobilize, enthuse and support people in carrying out the detailed work of the various phases.

- The performance assessment and feedback to the team.

All of these will need to be set in the context of the culture in which the changes are taking place and the initial focus should be on the needs of the senior team.

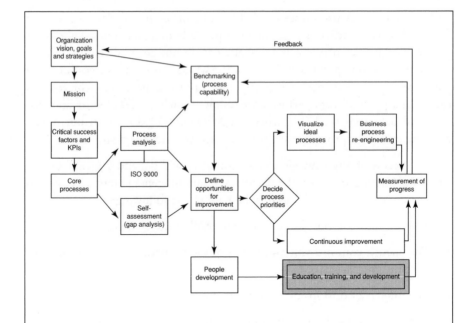

People – communications and training _____

Chapter 11

People – communications and training _____

Key points

People's attitudes and behaviour can be influenced by communication, and the essence of changing attitudes is to gain acceptance through excellent communication processes.

The strategy and changes to be brought about must be clearly and directly communicated from top management to all staff/employees. People must know when and how they will be involved, what the processes are and the successes and benefits achieved. The first line supervision has important roles in communicating the key messages and overcoming resistance to change, and these are identified in the chapter.

Good leadership is mostly about good communications, the skills of which can be learned through training, but must be acquired through practice.

All training should occur in an improvement cycle of: ensure training is part of the policy and strategy, allocate responsibilities, define objectives, establish training organization, specify needs, prepare programmes and materials, implement and monitor, assess results, and review effectiveness. These phases are dealt with in this chapter together with issues concerning follow-up and consultancy.

Communicating strategy and change

People's attitudes and behaviour clearly can be influenced by communication; one has to look only at the media or advertising to understand this. The essence of changing behaviour in business is to gain acceptance for the need to change, and for this to happen, it is essential to provide relevant information, convey good practices, and generate interest, ideas and awareness through excellent communication processes. This is possibly the most neglected part of many organizations' operations, yet failure to communicate effectively creates unnecessary problems resulting in confusion, loss of interest and eventually in poor performance through apparent lack of guidance and stimulus.

A strategy developed by top management for the direction of the business/organization is only half the battle. An early implementation step must be the clear widespread communication of the strategy. This will require direct and clear communication from the top management to all staff and employees, to explain the need to focus on processes. Everyone will need to know their roles in understanding processes and improving their performance.

Components of this communication will include the:

- concept of total organizational excellence,

- importance of understanding business processes,

- need for improvement,

- approach which will be taken,

- individual and process group responsibilities,

- principles of process measurement.

The system for disseminating the message may cover all the conventional communication methods of seminars, departmental meetings, posters, newsletters, intranet, etc. First line supervision will need to review the ideas directly with all their staff, and a set of questions and answers may be suitably pre-prepared in support.

Once people understand the strategy, the management must establish the infrastructure. The required level of individual commitment is likely to be achieved, however, only if everyone understands the aims, the role they must play, and how they can implement process improvements. For this understanding a constant flow of information is necessary, including:

- when and how individuals will be involved,

- what the process involves,

- the successes and benefits achieved.

The most effective means of developing the personnel commitment required is to ensure people know what is going on. Otherwise, they will feel left out which in turn will lead to resentment and undermining of the whole process. The first line of supervision again has an important part to play in ensuring key messages are communicated and in building teams by demonstrating everyone's involvement and commitment. In the Larkins' excellent book

Communicating Change, the authors refer to three 'facts' regarding the best ways to communicate change to employees:

1 Communicate directly to supervisors (first-line).

2 Use face-to-face communication.

3 Communicate relative performance of the local work area.

The language used at the 'coal-face' will need attention in many organizations. Reducing the complexity and jargon in the written and spoken communications will facilitate comprehension. When written business communications cannot be read or understood easily, they receive only cursory glances, rather than the detailed study they require. **Simplify and shorten** must be the guiding principles. The communication model illustrated in Figure 11.1 indicates the potential for problems through environmental distractions, mismatches between sender and receiver (or more correctly, decoder) in terms of attitudes – towards the information and each other – vocabulary, time pressures, etc.

All levels of management should introduce and stress 'open' methods of communication by maintaining open offices, being accessible to staff/employees, and getting involved in day-to-day interactions and the detailed processes. This will lay the foundation for improved interactions **between** staff and employees, which is so essential for information flow and process improvement. Opening these lines of communication may involve the confrontation of many barriers and much resistance. Training and the supervisors'/management's behaviour should be geared to helping people accept responsibility for their own behaviour that often creates the barriers, and for breaking the barriers down by concentrating on the process rather than 'departmental' needs.

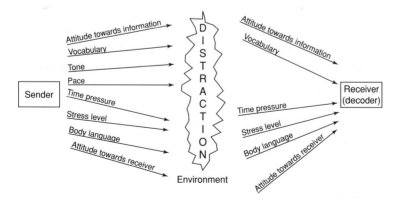

Figure 11.1 Communication model

Resistance to change will always occur and is to be expected. Again, the first line management must be trained to help people deal with it. This requires an understanding of the dynamics of change and the support necessary – not an obsession with forcing people to change. Opening up the lines of communication through a previously closed system, publicizing people's efforts to change and their results will aid the process. Change can be – even should be – exciting if employees start to share their development, growth, suggestions, and questions. Management must encourage and participate in this by creating the most appropriate communication systems.

Communicating the strategic message

People in most organizations designate themselves into one of four 'audience' groups, often each with particular general attitudes.

- Senior management – who should see process management as an opportunity, both for the organization and themselves.

- Middle management – who may see process management as another burden without any benefits, and who may perceive a vested interest in the status quo.

- Supervisors (first line or junior management) who may see process management as another 'flavour of the period' or campaign, and who may respond by trying to keep heads down so that it will pass over.

- Other employees who may not care, so long as they still have jobs and get paid, but these people must be the custodians of the delivery of the products or services to the customer and own that responsibility.

The key medium for motivating the employees and gaining their commitment is face-to-face communication and **visible** management commitment. Much is written and spoken about leadership, but it is mainly about communication. If people are good leaders they are invariably good communicators. Leadership is a human interaction, which depends on the communications between the leaders and the followers. It calls for many skills which can be **learned** from education and training but must be **acquired** through practice.

Types of communication

It may be useful to contemplate why people learn. They do so for several reasons, some of which are:

- self-betterment

- self-preservation

- need for responsibility

- saving time or effort

- sense of achievement

- pride in work

- curiosity

So communication and training can be a powerful stimulus to personal development at the workplace, as well as achieving improvements for the organization. This may be useful in the selection of the appropriate method(s) of communication, the principal ones being:

- **Verbal communication** either between individuals or groups using direct or indirect methods, such as public address and other broadcasting systems and recordings.

- **Written communication** in the form of notices, bulletins, information sheets, reports and recommendations.

- **Visual communication** such as posters, films, video tapes, exhibitions, demonstrations, displays and other promotional features. Some of these also involve verbal communication.

- **Example** through the way people conduct themselves and adhere to established working codes and systems; through their effectiveness as communicators and ability to 'sell' good practices.

The characteristics of each of these methods should be carefully examined before they are used in communicating the messages.

1 Direct verbal communication

The requirements of this method are for:
- careful preparation,
- good individual communication and presentation skills,
- a broad knowledge of the subject matter,
- ability to control and answer questions or seek answers,
- credibility with the audience or group,
- the encouragement of participation and involvement.

Its strengths are:
- direct impact on individuals or the group,
- permits assessment of reactions and allows discussion or presentations to be modified accordingly,

- permits use of plain words easily understood by groups or individuals,
- audience can ask questions and get answers,
- permits presenter to check assimilation through, for example, asking questions,
- allows for reiteration, recapitulation and special emphasis as necessary,
- 'personalizes' improvement,
- aids the process of participation and involvement,
- can secure commitment from groups and individuals.

But it has limitations; it:
- depends upon the individuals' ability to communicate effectively,
- uses only **one** of the senses through which people acquire knowledge,
- requires time to prepare carefully in proportion to the complexity of the subject,
- does not guarantee uniformity of content and understanding between groups unless based on a common agenda,
- is time-consuming and usually most effective for small groups.

The art of speaking to people effectively should be learned and practised at all levels of the communication framework.

2 Indirect verbal communication

This method is limited in its effectiveness in communicating. It suffers from many deficiencies. For example, many internal broadcasting systems are periodically overwhelmed by noise so there is no guarantee that the message has been even received – much less understood. It is often difficult to check that everyone has heard the message and checking understanding is often impractical. Furthermore, it is inflexible and cannot adapt easily to individual requirements.

3 Written communication

The requirements of written methods are:
- ability to express the message in words clearly and concisely,
- ability to make words interesting to read,
- ability to say exactly what is meant, unambiguously,
- sense of 'timing' – good administrative arrangements for circulation,
- awareness of limitations and deployment only in appropriate situations.

Its strengths are:
- same message goes to everyone,
- speedy,
- careful timing of 'release' can ensure that message is received by everyone at the same time,

- useful for dealing with large numbers in a short span of time,
- useful for giving 'non-critical' information,
- can be circulated by a number of routes simultaneously,
- helpful – and sometimes essential – in backing up verbal communication, particularly if subject is rather complex,
- regularizes and records actions, procedures, systems, rules, etc.,
- usual form for submitting reports and recommendations.

The limitations are:
- not everyone chooses to – or can – read; therefore no guarantee that message has 'got through',
- written words may mean different things to different people according to vocabulary,
- words may be ambiguous and create confusion and misunderstanding,
- no opportunity for clarification; cannot easily ask questions, get replies or discuss,
- difficult to convey relative importance and emphasis and give topic 'light and shade',
- lacks animation, de-personalizes communication processes and reduces opportunity for personal contact,
- reduces sense of involvement and precludes exchanges of information and views.

4   Visual communication

People learn through their senses but by far the highest percentage of what they take in is through sight. It is estimated that the five senses contribute to the learning process as follows:

sight (visible)	75%
hearing (audible)	13%
feeling (tactile)	6%
smell (olfactory)	3%
taste (gustatory)	3%

This clearly means that visible methods of communication can be extremely successful, especially when combined with other methods, such as verbal in the form of films, video, audiovisual presentations, or demonstrations. It also has implications for training sessions and group discussions in which visual aids should be used as liberally and dramatically as possible. The use of simple flip charts and felt-tip pens, which allow everyone to communicate on an equal footing, can be a very effective visual aid to most discussions.

5 **Communication by example**

Showing videos, displaying posters, discussion groups, speaking or writing are not the only ways of communicating. Personal example is a powerful medium for getting across the messages. This can be done by:

- people's general positive attitude and alertness;
- the way people conduct themselves at the work place;
- adherence to rules, procedures, systems, standard operations, and practices;
- standards of housekeeping and hygiene;
- the way in which people are inclined to help others appreciate and avoid potential problems;
- people learning how to relate to, communicate with and influence others to gain their commitment;
- the way in which people exude enthusiasm, pride, and confidence in themselves and the organization for which they work.

Written words, however well written, have no value unless backed by appropriate management behaviour. This is a most important means of communication which should proceed with continual emphasis on employee involvement and participation.

If people do have to be corrected at work it is important for managers and supervisors to remember that the objective is to help staff to understand and identify with their process problems and prevent recurrence of mistakes. It is necessary to be objective always and supervision should try to find out **what** has gone wrong rather than **who** has gone wrong. This will lead to reliance on facts rather than opinions and 'hunches'. Managers and first line supervisors should be trained to create a good atmosphere, be patient, listen, and be prepared to respond to other ideas and initiatives.

Communication skills

If any organization is to succeed, it is important for all managers, supervisors and staff to recognize the value and influence of good communication. Moreover, they must learn the characteristics of the various methods of communication and select the one most appropriate for the situation. This process is very much accelerated by an appreciation of how people learn to assimilate knowledge – since this will encourage people to make use of all the senses when communicating.

All communication and training exercises must be planned like a military operation, leaving nothing to chance. When dealing with people, managers and first line supervisors must be sensitive to the effect they have. They may need to be trained to communicate with people in a way that will help everyone feel more capable, more necessary and more worthwhile.

Effective communication is a two-way exercise. We were designed with two ears and one tongue. Could it be we were meant to listen twice as much as we speak? Listening with attention, interest, and courtesy and carrying on listening does not come naturally to many people and they may need to be reminded, in their training, teamwork, or counselling activities, to listen 'openly'.

Communicating in teams

The starting point for any improvement team is often brainstorming, during which a team member records all the ideas on a chart – possibly as a cause-and-effect diagram (see Chapter 9). The purpose and rules of brainstorming are directed at achieving agreement on action plans. The agreement to be reached among team members with regard to action plans will force the consideration of all aspects of a problem, and will make everyone alert to possible objections to the chosen courses of action. A useful device along this path to consensus of opinion is to allow the team some 'thinking time' before, during and after brainstorming sessions.

If conflict occurs, as is inevitable in any teamwork, it must be managed so that it assists rather than hinders the team to achieve its objectives. The team leader has an important role in such situations and it will help to:

- recognize all contributors of ideas, not just those whose ideas are used,

- stress that both the organization **and** the individuals benefit, if improvements are made,

- clarify what is expected of each team member, in terms of the common goals **and** individually assigned tasks,

- mediate when dominating members cause others to feel inadequate or suppressed,

- remind members who make personal references that the group is in existence to reach agreement and find solutions,

- endorse the positive traits of members, such as co-operation, openness, listening, contribution, etc.,

- discourage criticism, defensiveness, aggressiveness, closed-mindedness, interrupting, etc.

Talking to people

If anyone has to speak to a group of people, they should use the following 'check-list' as an aid to structuring and presenting the talk:

1   What is the **objective** of the talk – what has to be conveyed?

2   What **key points** must be included to achieve the objective?

3   In what **sequence** should these be arranged for maximum impact and smooth 'flow'?

4   Who will comprise the **audience**? What is their occupation, status level, experience, etc.? And how many will there be?

5   What **method** of presentation will be most effective for the particular purpose to be achieved and audience to be addressed? For example, will it be a talk or a discussion?

6   Will **visual aids** be required and if so how will they be used? The saying 'one picture is worth 1000 words' should remind people of the value of illustrating talks or discussions.

7   How much **time** is available and how will it be allocated?

8   In which **location** is the talk to be given? Is it suitable? Is it correctly appointed?

9   How will **assimilation** of the message be checked? Will some form of test and/or questions be used to check **understanding?**

10 What **follow-up** is planned to reinforce the message and ensure **implementation?**

In presenting a case or point of view in a meeting, people will need to be shown how to perform to maximum effect. The key points here are to encourage presenters to:

*   try, by taking a little time and by making introductions, to take the stress out of any 'negotiating' situations and create the right 'atmosphere';

*   speak slowly, simply and with variation in pitch and tone, using pauses to aid assimilation, and avoiding jargon and clichés;

*   present points of view logically and with clarity and precision so that all understand the substance of the case;

*   interest the audience through the content and presentation style, involving them whenever possible;

- keep an open mind and an open ear – listen carefully to what other people are saying;

- keep alert and try to read the 'language' of the situation;

- be flexible in outlook – adapt to new situations and take advantage of new avenues of approach;

- build on the best points of any particular case and diminish the weaknesses,

- make constructive proposals but prevent people from being 'cornered' inextricably;

- maintain discipline over the team or group to control discussions, cover the agenda, keep to the point, and keep cool!

- keep the objective in mind throughout, and do not try to cover too many points in one go;

- 'manage' the time of the meeting;

- be sure that both sides **understand** what has been agreed before the meeting closes;

- communicate outcomes quickly and accurately to all interested parties.

Training in presentation skills at all levels is never a waste of resources.

Reports and writing

There are some general ground-rules for effective presentation of written material and reports. A good report should be readable, interesting, informative, well presented and be no longer than is necessary. Recipients are likely to be busy people and will welcome a concise report from which they can grasp the essentials. Therefore, when constructing a report the following questions should be considered:

- Why has the report been requested?

- What are the terms of reference and objectives?

- What messages need to be conveyed?

- What type of information is required? For example, is it factual information based on observation or research, conclusions drawn from facts, or recommendations as to future courses of action?

- Is it arranged in a logical sequence, such as:

    Title

Contents
Introduction or terms of reference
Summary
Data and information
Analysis/discussion
Conclusions and recommendations
Action plans
Supplementary information (e.g. appendices)

- Will the subject and presentation capture the interest of the reader?

- Is it as short and easy to read as possible?

- Can some of the text be explained in diagrammatic form where this will aid understanding?

- Is it intelligible to the reader?

- Is it free of 'shorthand' expressions, departmental 'code words' and technical jargon?

- Does it take account of minority views or dissenting opinions?

- Have factual information and personal comments been distinguished to provide a sufficiently objective report?

- Do the conclusions and recommendations match the requirements of the terms of reference? Do they follow logically from the information and analysis contained in the main body of the report?

Report writing requires a particular skill in expression and some experience to do well. These notes are only a few clues to the more obvious considerations and are not exhaustive. There are many full texts on this subject.

Flowcharts paint written 'pictures' of processes which any group of people can understand and use (see Chapter 5). Any process will have its own communication system, with its own separate and distinct flow. This should be recognized, flow-charted and understood as an integral part of the process, and it may be superimposed on the flow chart of activities.

Leading discussions

For those who have to lead discussions, including committee chairmen and leaders of improvement teams, this plan may help:

Make an outline: Determine the objectives and what is to be covered.
Decide the 'key points' for discussion and how much time there will be available.

	Select areas of priority from 'key points'; relative to the time available. Plan the session in the imagination.
Plan the approach:	Decide how the topic will be introduced. Move the discussion along from one point to the next. Determine how people will be brought into the discussion.
Plan the physical arrangements:	Make sure everyone can be made comfortable. Ensure that everyone can see and hear. Check the heating, ventilation, lighting, etc. Make sure all the necessary visual aids are available and that they are all serviceable.
Introduce the session:	Review the background to the discussion. Announce the topic briefly and concisely and emphasize the relevance to the background. Explain how the discussion should proceed and gain commitment to this approach. Lead into the discussion smoothly and logically.
Control the discussion	Encourage participation; draw out ideas by asking questions, encouraging the exchange of views and opinions. Encourage the reluctant contributors and prevent monopolization by the more vocal members. Distribute the questions evenly and avoid bias. Keep the discussions to the point and always moving forward. Stimulate thought and discussion if necessary. Handle irrelevancies tactfully. Summarize frequently, particularly at key stages of the dicussion.
Summarize the discussion and document	Summarize the various outputs from the discussion; the ideas and experiences, etc., which came to light. Restate the objective. Arrive at conclusions or solutions and restate as an achievement. Give credit for effective contributions. Issue a written statement of the main points and conclusions, with action plans if appropriate, as soon as possible.

Chairing meetings

Many factors make a good chairman of a meeting. The following points should help people charged with such tasks to improve their performance:

1   Initiate introductions if members of the group do not know each other.

2   Create a good 'atmosphere' in which the meeting can proceed smoothly. Remain calm and cool. Be impartial.

3   State the intentions of the meeting and 'ground rules' for its conduct. Gain tacit acceptance of this 'methodology'.

4   Maintain a good-natured discipline over the meeting; control casual conversations and emotional outbursts. Involve all members present.

5   Be courteous, patient and understanding. Not all individuals are articulate – help them to overcome their problems of self-expression so that their views, ideas and experience are not disregarded.

6   Keep an eye on the clock and make sure that time is allocated usefully.

7   Be flexible. If the meeting is getting 'bogged down', allow discussions a little freedom but do not lose sight of the objectives or the time in the process. A little humour applied at the right time might help to revive a flagging meeting. Adjust the speed of debate up or down according to the situation.

8   Cover the agenda and achieve the intentions. Make interim summaries. Indicate progress; re-focus off-track discussions; highlight or confirm important points; clarify and restate points which are not clear.

9   Involve the group as much as possible; generate discussion to reach solutions and decide options, according to requirements.

10   Summarize the consensus views.

In planning all communication activity, answering **six basic questions** will help the process get of to a flying start.

- **Why** should we communicate?

- **What** should we communicate?

- **Who** should we communicate with?

- **How** should we communicate?

- **When** should we communicate?

- **Where** should we communicate?

The more that is known and the better is the information gathered in response to these questions, the more effective will be the communications.

It's Friday – it must be training

It is the author's belief that training people is the single most important factor in actually improving performance. For training to be effective, however, it must be planned in a systematic and objective manner. Training must be continuous to meet not only changes in technology, but also changes involving the environment in which an organization operates, its structure, and perhaps most important of all the people who work there.

Training activities can be considered in the form of a cycle of improvement (Figure 11.2), the elements of which are:

Ensure training is linked to policy and strategy

Every organization should define clearly its policies and strategies (Chapters 2 and 3) which should contain principles and goals to provide a framework within which training activities will be planned and operated.

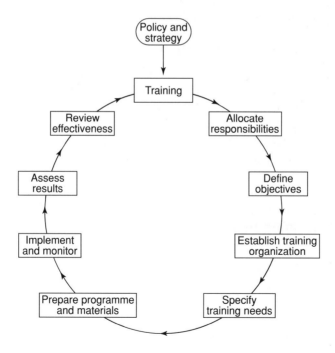

Figure 11.2 The training circle

Allocate responsibilities for training

Training should be the responsibility of line management, but there are also important roles for the HR manager and his/her colleagues.

Define training objectives

The following questions are useful first steps when identifying training objectives:

- How are the customer requirements and business objectives transmitted through the organization?

- Which areas demand improved performance?

- What changes are planned for the future?

When attempting to set training objectives, three essential requirements must be met:

    i)   senior management must ensure that objectives are clarified and priorities set,

    ii)   defined objectives must be realistic and attainable,

    iii) the main problems should be identified for all areas in the organization; the large organization may find it necessary to promote a phased plan to identify these problems.

Establish training organization

In every organization, the overall responsibility for seeing that training is properly organized must be assumed by one or more designated senior executives. All managers have a responsibility for ensuring that personnel reporting to them are properly trained and competent in their job – this responsibility should be written into every manager's job description. The question of whether line management requires specialized help should be answered when objectives have been identified. It is often necessary to use specialists who may be internal or external to the organization.

Specify training needs

The next step in the cycle is to assess and clarify specific training needs. The following questions need to be answered:

- Who needs to be trained?

- What competencies are required?

- How long will training take?

- What are the expected benefits?

- Is the training need urgent?
- How many people are involved?
- Who will undertake the actual training?
- What resources are needed, e.g. money, people, equipment, accommodation, outside resources?

Prepare training programmes and materials

Line management should participate in the creation of draft programmes and, as they will often need to create the training programmes themselves, they should retain the final responsibility for what is implemented.

Training programmes should include:

a)   the training objectives expressed in terms of the desired behaviour,

b)   the actual training content,

c)   the methods to be adopted,

d)   who is responsible for the various sections of the programme.

Implement and monitor training

The effective implementation of training programmes demands considerable commitment and adjustment by the trainers and trainees alike. Training is a progressive process which must take into account any learning problems of the trainees.

Access the results

In order to determine whether further training is required, line management should themselves review performance when training is completed. However good the training may be, if it is not valued and built upon by managers and supervisors its effect can be severely reduced.

Review the overall effectiveness

Senior management require a system whereby decisions are taken at regular fixed intervals on:

- the training policy and strategy
- the training objectives
- the training organization

Even if the policy remains constant, there is a continuing need to ensure that new training objectives are set either to promote work changes or to raise the standards already achieved.

The purpose of management system audits and reviews is to assess the effectiveness of an organization's activities. Clearly, adequate and refresher training in these methods is essential if such checks are to be realistic and effective. Audits and reviews can provide useful information for the identification of changing training needs.

The training organization should similarly be reviewed in the light of the new objectives, and here again it is essential to aim at continuous improvement. Training must never be allowed to become static, and the effectiveness of the organization's training programmes and methods must be assessed and reviewed systematically.

Follow-up and consulting

To be successful training must be followed up and this can take many forms, but the managers must provide the lead through the design of appropriate improvement projects and 'surgery' workshops.

In introducing methods of process improvement, for example, the most satisfactory strategy is to start small and build up a bank of knowledge and experience. Improvements in one or two areas of the organization's operations, using this approach, will quickly establish the techniques as reliable methods.

The author and his colleagues have found that a successful formula is the in-company-training course plus follow-up workshops. Typically a 1–2 day seminar on process management is followed within a few weeks by a ½–1 day workshop at which participants on the initial training course present the results of their efforts to improve processes and use the various methods. The presentations and specific implementation problems may be discussed. A series of such workshops will add continually to the follow-up and can be used to initiate process improvement teams. Wider company presence and activities should be encouraged by the follow-up activities.

It will usually be found that external help is required to introduce and establish the necessary components of total organizational excellence. Again the author has lost count of the number of occasions on which, following a presentation in a board room of a company, he has been challenged by the sponsoring director or manager, who claims that he has been repeating for years all the points and suggestions that he has just heard the 'outsider' pronounce with such authority. The difference, of course, is that the company is now going to do something about it. It is not clear why this happens, perhaps it is about presentation skills, perhaps the prophet is never accepted in his own land, but the fact remains that external advice is often heeded and, therefore, needed.

If external consultants and trainers are used, and both large and small companies buy in skills in this field, it is essential to carefully select and control the counselling. The dangers of not doing so are:

- the creation of a 'system' which is not operational,

- the 'not-invented-here' effect of buying in someone else's methods,

- a mismatch of the consultant's approach and the unique requirements of the company, its style, operations and the business environment in which it lives.

Any of these will render the 'system' unworkable.

The fact that a consultant is 'known to operate' in an industry is not quite the same as the consultant knowing what the requirements of the industry or company are and how they differ from those of other industries. There are many consultancy organizations which have failed to provide good advice because they did not possess the depth of knowledge of both management issues and their application in special processes. For example, the application of statistical process control (SPC) methods in continuous polymer production is not simply a question of changing widgets for polyolefins, it requires a fundamentally different approach, which derives from the direct experience of the consultant and his/her understanding of the nature of the process or industry involved.

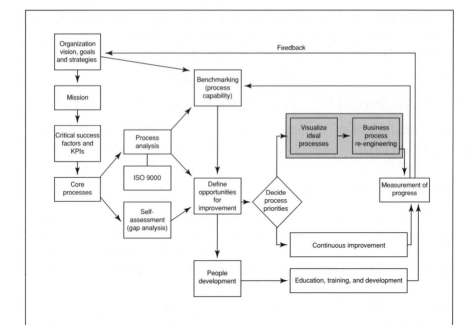

Re-visioning and business process re-design _____

Chapter 12

Re-visioning and business process re-design

Key points

Total organizational excellence requires not only commitment but also a competence in the mechanics of process management.

Successful practitioners of business process re-engineering (BPR) have made striking improvements in customer satisfaction and productivity in short periods of time by following some simple steps which are explained in this chapter. Crucial early stages will involve establishment of the current process structure, collecting performance measurement information, teamwork, systems and training.

Change must create something which did not exist before, namely a 'learning organization' capable of adapting to a changing competitive environment. When processes or even the whole business needs to be re-engineered, the radical change may, probably **will**, not be readily accepted. This chapter will provide guidance and support to those engaged in this most difficult of sciences/arts. It describes the typical phases of a project and considers the issues of managing the changes, and the roles of information technology in supporting BPR.

Re-engineering the organization?

When it has been recognized that a major business process requires radical re-assessment, business process re-engineering or re-design (BPR) methods are appropriate. In their book *Re-Engineering the Corporation*, Hammer and Champy talked about re-inventing the nature of work, 'starting again – re-inventing our corporations from top to bottom.' BPR was launched on a wave of organizations needing to completely re-think how and why they do what they do in order to cope with the ever-changing world, particularly the development of IT solutions.

The reality is, of course, that many processes in many organizations are very good and do not need re-engineering, re-designing or re-inventing, not for a while anyway. These processes should be subjected to a regime of continuous improvement (Chapter 13), at least until we have dealt with the very poorly performing processes that clearly do need radical review.

Some businesses and industries more than others have been through some pretty hefty changes – technological, political, financial and/or cultural. Customers of these organizations may be changing and demanding certain new relationships. Companies are finding leaner competitors encroaching into their market place, increased competition from other countries where costs are lower, and start-up competitors which do not share the same high bureaucracy and formal structures.

Enabling an organization, whether in the public or private sector, to be capable of meeting these changes is not a case of working harder but working differently. There have been many publicized BPR success stories and, equally, there have been some abject failures. In some cases, radical changes to major business processes have brought corresponding radical improvements in productivity. However, knowing how to reap such benefit, or indeed knowing if and how to apply BPR, has proved difficult for some organizations.

The concept of BPR was introduced to the world via two articles that described the radical changes to business processes being performed by a handful of western businesses. These were also among the first to embark on TQM initiatives in the 1980s, and included Xerox, Ford, AT&T, Baxter Healthcare, and Hewlett-Packard.

Many companies adopted TQM initiatives in the 1980s hoping to win back business lost to Japanese competition. When Ford benchmarked Mazda's accounts payable department, however, they discovered a business process being run by five people, compared to Ford's 500. Even with the difference in scale of the two companies, this still demonstrated the relative inefficiency of Ford's accounts payable process. At Xerox, taking a customer's perspective of the company identified the need to develop systems rather than stand-alone products, which highlighted Xerox's own inefficient office systems.

Both Ford and Xerox realized that incremental improvement alone was not enough. They had developed high infrastructure costs and bureaucracies that made them relatively unresponsive to customer service. Focusing on internal customer–supplier interfaces improved quality, but preserved the current process structure and they could not hope to achieve in a few years what had taken the Japanese 30 years. To achieve the necessary improvements required a radical rethinking and redesign of these processes.

What was being applied by organizations such as Ford and Xerox was **discontinuous improvement.** In order to respond to the competitive threats of

Canon and Honda, Xerox and Ford needed TQM to catch up; but to get ahead they felt they required radical breakthroughs in performance. Central to these breakthrough improvements was information technology (IT).

Information technology as a driver for BPR

BPR is often based on new possibilities for breakthrough performance provided by the emergence of new enabling technologies. The most important of these, the one that is the nominal ingredient in many BPR recipes, is IT.

Explosive advances in IT have enabled the dissemination, analysis, and use of information from and to customers and suppliers and within enterprises, in new ways and in time frames that impact processes, organization designs and strategic competencies. Computer networks, open systems, client-server architecture, groupware, and electronic data interchange (EDI) have opened up the possibilities for the integrated automation of business processes. Neural networks, enterprise analyser approaches, computer-assisted software engineering, and object-oriented programming now facilitate systems design around office processes.

The pace of change has, of course, been enormous and IT systems unavailable just ten to fifteen years ago have enabled sweeping changes in business process improvement, particularly in office systems. Just as statistical process control (SPC) enabled manufacturing processes to be improved by controlling variation and improving efficiency, so IT is enabling non-manufacturing processes to be fundamentally restructured.

IT in itself, however, did not offer all the answers, automation frequently being claimed not to produce the gains expected. Many companies putting in major new computer systems have achieved only the automation of existing processes. Frequently, different functions within the same organization have systems that are incompatible with each other. Locked into traditional functional structures, managers have spent large amounts on IT systems that have not been used cross-functionally. Yet it is in this cross-functional area that the big improvement gains through IT are to be made. Once a process view is taken to designing and installing an IT system, it becomes possible to automate cross-functional, cross-divisional, even cross-company processes.

What is BPR and what does it do?

There are almost as many definitions of BPR as there are of TQM! However, most of them boil down to the same substance – the fundamental rethink and radical re-design of a business process, its structure, and associated management systems, to deliver major or step improvements in performance (which may be in process, customer, or business performance terms).

Of course, BPR and TQM programmes are complimentary under the umbrella of process management. The continuous and step change improvements must live side by side – when does continuous change become a step change anyway? There has been over the years much debate, including some involving the author, about this issue. Whether it gets resolved is not usually the concern of the organization facing today's uncertainties with the realization that 'business as usual' will not do and some major changes in the ways things are done are required.

Put into a strategic context, BPR is a means of aligning work processes with customer requirements in a dynamic, flexible way, in order to achieve long-term corporate objectives. This requires the involvement of customers and suppliers and thinking about future requirements. Indeed the secrets to re-designing a process successfully lie in thinking about how to reshape it for the future.

BPR then challenges managers to rethink their traditional methods of doing work and to commit to customer-focused processes. Many outstanding organizations have achieved and/or maintained their leadership through process re-engineering, especially where they found processes which were not customer focused. Companies using these techniques have reported significant bottom-line results, including better customer relations, reductions in cycle time to market, increased productivity, fewer defect/errors and increased profitability. BPR uses recognized methods for improving business results and questions the effectiveness of the traditional organizational structure. Defining, measuring, analysing and re-engineering work processes to improve customer satisfaction can pay off in many different ways.

For example, Motorola had set stretch goals of ten-fold improvement in defects and two-fold improvement in cycle time within five years. The time period was subsequently revised to three years and the now famous Six Sigma goal of 3.4 defects per million became a slogan for the company and probably one of the real drivers. These stretch goals represent a focus on discontinuous improvement and there are many examples of other companies that have made dramatic improvements following major organizational and process redesign as part of TQM initiatives, including approaches such as the 'clean sheet' design of a 'green field' plant around work cells and self-managed teams.

Most organizations have vertical functions: experts of similar backgrounds grouped together in a pool of knowledge and skills capable of completing any task in that discipline. This focus, however, fosters a vertical view and limits the organization's ability to operate effectively. Barriers to customer satisfaction evolve, resulting in unnecessary work, restricted sharing of resources, limited synergy between functions, delayed development time and no clear understanding of how one department's activities affect the total process of attaining customer satisfaction. Managers remain tied to managing singular functions, with rewards and incentives for their narrow missions, inhibiting a shared external customer perspective.

BPR breaks down these internal barriers and encourages the organization to work in cross-functional teams with a shared horizontal view of the business. As we have seen in earlier chapters it requires shifting the work focus from managing functions to managing processes. Process owners, accountable for the success of major cross-functional processes, are charged with ensuring that employees understand how their individual work processes affect customer satisfaction. The interdependence between one group's work and the next becomes quickly apparent when all understand who the customer is and the value they add to the entire process of satisfying that customer.

Processes for re-design

IT provided the means to achieve the breakthrough in process performance in some organizations. The inspiration, however, came from understanding both the current and potential processes. This required a more holistic view than that taken in traditional quality programmes, involving wholesale redesigns of the processes concerned.

Ford estimated a 20 per cent reduction in head count if it automated the existing processes in accounts payable. Taking an overall process perspective, Ford achieved a 75 per cent reduction in one department. Xerox took an organizational view and concentrated on the cross-functional processes to be re-engineered, radically changing the relationship between supplier and external customer.

Clearly, the larger the scope of the process, the greater and farther reaching are the consequences of the redesign. At a macro level, turning raw materials into a product used by a delighted customer is a process made up of subsets of smaller processes. The aim of the overall process is to add value to the raw materials. Taking a holistic view of the process makes it possible to identify non-value-adding elements and remove them. It enables people to question why things are done, and to determine what should be done.

Some of the re-engineering literature advised starting with a blank sheet of paper and redesigning the process anew. The problems inherent in this approach are:

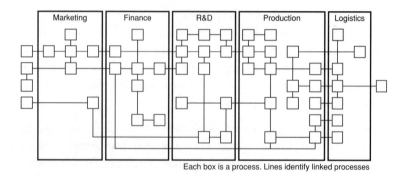

Each box is a process. Lines identify linked processes

Figure 12.1 Simplified process map

1 the danger of designing another inefficient system, and

2 not appreciating the scope of the problem.

Therefore, the author and his colleagues recommend a thorough understanding of current processes before embarking on a re-engineering project.

Current processes can be understood and documented by process mapping and flowcharting. As processes are documented, their inter-relationships become clear and a map of the organization emerges. Figure 12.1 shows a much simplified process map. As the aim of BPR is to make discontinuous, major improvements, this invariably means organizational change, the extent of which depends on the scope of the process re-engineered.

Taking the organization depicted in Figure 12.1 as an example, if the decision is made to redesign the processes in finance, the effect may be that in Figure 12.2a: eight individual processes have become three. There has been no organizational effect on the processes in the other functions, but finance has been completely restructured. In Figure 12.2b, a chain of processes, crossing all the functions has been re-engineered. The effect has been the loss of redundant processes and possibly many heads, but much of the organization has been unaffected. Figure 12.2c shows the organization after a thorough re-engineering of all its processes. Some elements may remain the same, but the effect is organization-wide.

Whatever the scope of the redesign, head count is not the only change. When work processes are altered, the way people work alters. Figures 12.1 and 12.2 show an organization's functional departments with processes running through them. These are the handful of core processes that make up what an organization does (see Figure 12.3) and in many organizations these would benefit from re-engineering to improve added value output and efficiency.

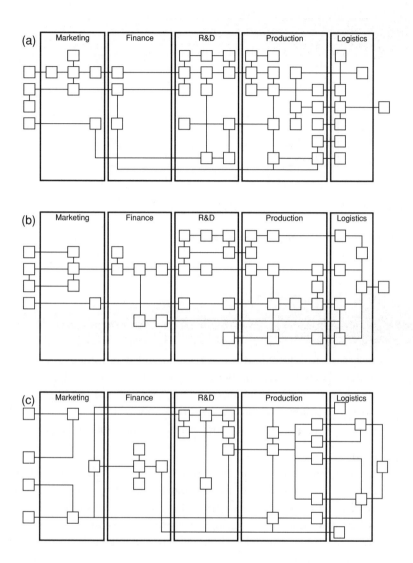

Figure 12.2 (a) Process redesign in finance, (b) cross-functional process redesign, (c) organizational process redesign.

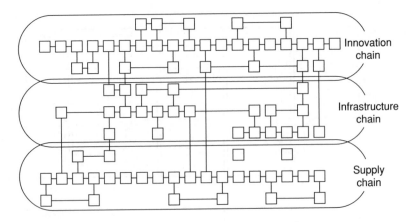

Figure 12.3   Process organization

Focus on results

BPR is not intended to preserve the status quo, but to fundamentally and radically change what is done; it is *dynamic*. Therefore, it is essential for a BPR effort to focus on required customers which will determine the scope of the BPR exercise. A simple requirement may be a 30 per cent reduction in costs or a reduction in delivery time of two days. These would imply projects with relatively narrow scope, which are essentially inwardly focused and probably involve only one department; for example, the finance department in Figure 12.2a.

When Wal-Mart focused on satisfying customer needs as an outcome, it started a redesign that not only totally changed the way it replenished inventory, but also made this the centrepiece of its competitive strategy. The system put in place was radical, and required tremendous vision. In ten years, Wal-Mart grew from being a small niche retailer to the largest and most profitable retailer in the world.

Focusing on results rather than just activities can make the difference between success and failure in change programmes. The measures used, however, are crucial. At every level of re-engineering, a focus on results gives direction and measurability; whether it be cost reduction, head count reduction, increase in efficiency, customer focus, identification of core processes and non-value-adding components, or strategic alignment of business processes. Benchmarking is a powerful tool for BPR and is the trigger for many BPR projects, as in Ford's accounts payable process. As shown in Chapter 8, the value of benchmarking does not lie in what can be copied, but in its ability to identify goals. If used well, benchmarking can shape strategy and identify potential competitive advantage.

The redesign process

Central to BPR is an objective overview of the processes to be redesigned. Whereas information needs to be obtained from the people directly involved in those processes, it is never initiated by them. Even at its lowest level, BPR has a top-down approach and most BPR efforts, therefore, take the form of a project. There are numerous methodologies proposed, but all share common elements. Typically, the project takes the form of seven phases, shown in Figure 12.4.

1 Discover

This involves firstly identifying a problem or unacceptable outcome, followed by determining the desired outcome. This usually requires an assessment of the business need and will certainly include determining the processes involved, including the scope, identifying process customers and their requirements, and establishing effectiveness measurements.

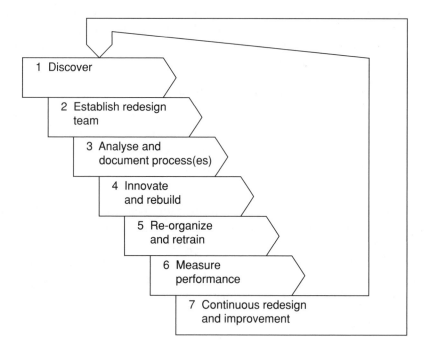

Figure 12.4   The seven phases of BPR

2 Establish redesign team

Any organization, even a small company, is a complex system. There are customers, suppliers, employees, functions, processes, resources, finances, etc. and many large organizations are incomprehensible – no one person can easily get a clear picture of all the separate components. Critical to the success of the redesign is the make-up of a redesign team.

The team should comprise as a minimum the following:

- senior manager as sponsor,
- steering committee of senior managers to oversee overall re-engineering strategy,
- process owner,
- team leader,
- redesign team members.

It is generally recommended that the redesign team have between five and ten people; represent the scope of the process (that is, if the process to be re-engineered is cross-functional, so is the team); only work on one redesign at a time; and is made up of both insiders and outsiders. Insiders are people currently working within the process concerned who help gain credibility with co-workers. Outsiders are people from outside the organization who bring objectivity and can ask the searching questions necessary for the creative aspects of the redesign. Many companies use consultants for this purpose.

3 Analyse and document process(es)

Making visible the invisible, documenting the process(es) through mapping and/or flow-charting is the first crucial step that helps an organization see the way work really is done and not the way one thinks or believes it is done. Seeing the process as it is provides a baseline from which to measure, analyse, test and improve.

Collecting supporting process data, including benchmarking information and IT possibilities, allows people to weigh the value each task adds to the total process, to rank and select areas for the greatest improvement, and to spot unnecessary work and points of unclear responsibility. Clarifying the root causes of problems, particularly those that cross department lines, safeguards against quick-fix remedies and assures proper corrective action, including the establishment of the right control systems.

4 Innovate and rebuild

In this phase the team rethink and redesign the new process, using the same process mapping technique, in an iterative approach involving all the stakeholders, including senior management. Approval, by all, of the action plan commits the organization to implementing the changes and following them through. The new process redesign is established in this phase.

5 Re-organize and re-train

This phase includes piloting the changes and validating their effectiveness. The new process structure and operation/system will probably lead to some re-organization, which may be necessary for reinforcement of the process strategy and to achieve the new levels of performance.

Training and/or re-training for the new technology and roles play a vital part in successful implementation. People need to be equipped to assess, re-engineer, and support – with the appropriate technology – the key processes that contribute to customer satisfaction and corporate objectives. Therefore, BPR efforts can involve substantial investment in training but they also require considerable top management support and commitment.

6 Measure performance

It is necessary to develop appropriate metrics for measuring the performance of the new process(es), sub-processes, activities, and tasks. These must be meaningful in terms of the inputs and outputs of the processes, and in terms of the customers of and suppliers to the process(es).

7 Continuous redesign and improvement

The project approach to BPR suggests a one-off approach. When the project is over, the team is disbanded, and business returns to normal, albeit a radically different normal. It is generally recommended that an organization does not attempt to re-engineer more than one major process at a time, because of the disruption and stress caused. Therefore, in major re-engineering efforts of more than one process, as one team is disbanded, another is formed to redesign yet another process. Considering that Ford took five years to redesign its accounts payable process, BPR on a large scale is clearly a long-term commitment.

In a rapidly changing, ever more competitive business environment, it is becoming more likely that companies will re-engineer one process after another. Once a process has been redesigned, continuous improvement of the new process by the team of people working in the process should become the norm.

BPR – the people and the leaders

For a company to focus on its core processes almost certainly requires an understanding of its core competencies. Moreover, core process redesign can channel an organization's competencies into an outcome that gives it strategic competitive advantage and the key element is visioning that outcome. Visioning the outcome may not be enough, however, since many companies 'vision' desired results without simultaneously 'visioning' the systems that are required to generate them. Without a clear vision of the systems, processes, methods, and approaches that will allow achievement of the desired results, dramatic improvement is frequently not obtained as the organization fails to align around a common tactical strategy. Such an 'operational vision' is lacking in many companies.

The fallout from BPR has profound impacts on the employees in any enterprise at every level – from executives to operators. In order for BPR to be successful, therefore, significant changes in organization design and enterprise culture are also often required. Unless the leaders of the enterprise are committed to undertake these changes, the BPR initiative will flounder. The difficulty is, of course, that organization design and culture change are much more difficult than modifying processes to take advantage of new IT.

While the enabling IT is often necessary and is clearly going to play a role in many BPR exercises, it is by no means sufficient, nor is it the most difficult hurdle on the path to success. Thanks to IT we can radically change the processes an organization operates and, hopefully, achieve dramatic improvements in performance. However, in any BPR project there will be considerable risk attached to building the information system that will support the new, redesigned processes. Information systems should be but rarely are described so that they are easy for people to understand.

While BPR is a distinct, short-term activity for a specific business function, the record indicates that BPR activities are most successful when they occur within the framework of a long-term thrust for excellence. Within the total organizational excellence framework, the BPR effort is more likely to find the process focus supportive workforce, organization design, and culture changes needed for its success.

Quality improvement is often positioned as a bottom-up activity. In some contrast, the total organizational excellence framework involves setting longer-term goals at the top, a self- assessment, and modifying the business as necessary to achieve the goals. Often, the modifications to the business required to achieve the goals are extensive and groundbreaking. The history of successful TQM thrusts in award-winning companies in Europe and the United States is replete with new organization designs, with flattened structures, and with empowering employees in the service of end customers. In many successful organizations, BPR has been an integral part of the culture – a process-driven culture dedicated to the ideals and concepts of TQM.

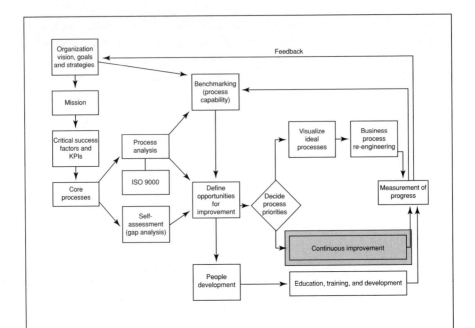

Continuous improvement _____

Chapter 13

Continuous improvement _____

Key points

Continuous improvement involves planning and operating processes, providing inputs, evaluating outputs, examining performance, and modifying processes and their inputs to achieve better performance. This chapter explains the three basic principles of continuous improvements: focus on the customer, understand the process and involve the people.

In Oakland's model for total quality management, the customer–supplier chains form the core which is surrounded by the hard management necessities of a good quality system, tools and teamwork. These need to work together to manage, measure and improve processes continually. Continuous improvement methods may be used to 'check' progress, in terms of commitment, strategies, teamwork, problems and results, and development towards excellence.

A structured approach to making improvements is provided by the **drive** model:

- **d**efine the problem,
- **r**eview the information,
- **i**nvestigate the problem,
- **v**erify the solution, and
- **e**xecute the change.

People working in a process must know whether it is capable of meeting the requirements, know whether it is actually doing so at any time, and make corrections when it is not. Simple SPC techniques are used, not only as a tool kit, but as a strategy for reducing variability, part of the continuous improvement approach.

The basic principles of continuous improvement

Continuous improvement is probably the most powerful concept to guide management. It is a term not well understood in many organizations, although that must begin to change if those organizations are to survive. The concept requires a systematic approach with the following components:

- *planning* the process and their inputs,

- *providing* the inputs,

- *operating* the processes,

- *evaluating* the outputs,

- *examining* the performance of the processes,

- *modifying* the processes and their inputs.

This system must be firmly tied to a continuous assessment of customer needs, and depends on a flow of ideas on how to make improvements, reduce variation, and generate greater customer satisfaction. It also requires a high level of commitment, and a sense of personal responsibility in those operating the processes.

The continuous improvement cycle of plan, do, check, act, ensures that the organization learns from results, standardizes what it does well in a documented management system, and improves operations and outputs from what it learns. But the emphasis must be that this is done in a planned, systematic, and conscientious way to create a climate – a way of life – that permeates the whole organization.

There are three basic principles of continuous improvement:

- focusing on the *customer,*

- understanding the *process,* and

- commitment from the *employees.*

1 Focusing on the customer

An organization must recognize, throughout its ranks, that the purpose of all work and all efforts to make improvements is to serve the customers better. This means that it must always know how well its outputs are performing, in the eyes of the customer, through measurement and feedback. The most important customers are the external ones, but the quality chains can break down at any point in the flows of work. Internal customers therefore must also be well served if the external ones are to be satisfied.

2 Understanding the process

In the successful operation of any process it is essential to understand what determines its performance and outputs. This means intense focus on the design and control of the inputs, working closely with suppliers, and understanding process flows to eliminate bottlenecks and reduce waste. If there is. one difference between management/supervision in the Far East and the West, it is that in the former management is closer to, and more involved in the processes. It is not possible to stand aside and manage in continuous improvement, it means that everyone has the determination to use their detailed knowledge of the processes and make improvements, and to use appropriate statistical methods to analyse and create action plans.

To begin to monitor and analyse any process, it is necessary first of all to identify what the process is, and what the inputs and outputs are. Many processes are easily understood and relate to known procedures, e.g. drilling a hole, compressing tablets, filling cans with paint, polymerizing a chemical using catalysts. Others are less easily identified, e.g. servicing a customer, delivering a lecture, storing a product in a warehouse, inputting to a computer. In many situations it can be extremely difficult to define the process. For example, if the process is inputting data into a computer terminal, it is vital to know if the scope of the process includes obtaining and refining the data, as well as inputting. Process definition is so important because the inputs and outputs change with the scope of the process.

Once the process is specified, the inputs and suppliers, outputs and customers can also be defined, together with the requirements at each of the interfaces. Some processes may produce primary and secondary outputs, such as a telephone call answered **and** a message delivered. If the requirements are not clarified or quantified, they are often assumed or estimated. Even if this does not lead to direct complaints, it will lead to waste – lost time, confusion – and perhaps lost customers. It is salutary for some suppliers of internal customers to realize that the latter can sometimes find new suppliers if their true requirements are not properly identified and/or repeatedly not met.

Prevention of failure in any transformation is possible only if the process definition, flow, inputs, and outputs are properly documented and agreed. The documentation of procedures will allow reliable data about the process itself to be collected, analysis to be performed, and action to be taken to improve the process and prevent failure or non-conformance with the requirements. The target in the operation of any process is the total avoidance of failure. If the idea of no-failures or error-free work is not adopted, at least as a target, then it certainly will never be achieved.

3 Commitment from the employees

Everyone in the organization, from top to bottom, from offices to technical service, from headquarters to local sites, must play their part. People are the source of ideas and innovation, and their expertise, experience, knowledge, and co-operation have to be harnessed to get those ideas implemented.

When people are treated like machines, work becomes uninteresting and unsatisfying. Under such conditions it is not possible to expect quality services and reliable products. The rates of absenteeism and of staff turnover are measures that can be used in determining the strengths and weaknesses, or management style and people's morale, in any company.

The first step is to convince everyone of their own role in the improvement regime. Employers and managers must of course take the lead, and the most senior executive has a personal responsibility. The degree of management's enthusiasm and drive will determine the ease with which the whole workforce is motivated.

Most of the work in any organization is done away from the immediate view of management and supervision, and often with individual discretion. If the co-operation of some or all of the people is absent, there is no way that managers will be able to cope with the chaos that will result. This principle is extremely important at the points where the processes 'touch' the outside customer. Every phase of these operations must be subject to continuous improvement, and for that everyone's co-operation is required.

Continuous improvement is the process by which greater customer satisfaction is achieved. Its adoption recognizes that quality is a moving target, but its operation actually results in quality.

Implementing teamwork for continuous improvement – the 'drive' model

The author and his colleagues have developed a model for a structured approach to continuous improvement in teams, the **drive** model. The mnemonic provides landmarks to keep the team on track and in the right direction:

- **D**efine – improvement opportunity.
  Output: written definition of the task and its success criteria.

- **R**eview – the information.
  Output: presentation of known data and action plan for further data.

- **I**nvestigate – the process.
  Output: documented proposals for improvement and action plans.

- **V**erify – the solution.
  Output: proposed improvements which meet success criteria.

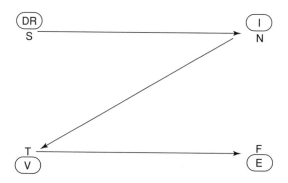

Figure 13.1 The drive model and MTBI-based problem solving

- Execute – the change.
  Output: task achieved and improved process documented.

The drive model fits well with the MBTI Z shaped problem solving approach (see Chapter 10). Figure 13.1 shows how the five stages relate to the S–N–T–F path.

The various stages are discussed in detail below. Some of the steps may be omitted if they are already answered or are clearly not relevant to a particular situation.

Define

At this stage the team is concerned with gaining a common understanding and agreement within the groups of the task that it faces, in terms of the improvement opportunity and the boundaries of the process or processes that contain it. It is necessary to generate at the outset a means of knowing when the team has succeeded. It is not concerned at this stage with solutions. The key steps are:

1 Interrogate the task. Typical questions:
   a) What is the brief?
   b) Is it understood?
   c) Is there agreement with it?
   d) Is it sufficiently explicit?
   e) Is it achievable?

   There may be a need for clarification with the 'sponsor' at this stage and possibly some redefinition of the task.

2 Understand the process:
   a) Which processes 'contain' the problem?
   b) What is wrong at present?
   c) Brainstorm – ideas for improvement.
   d) Perhaps draw a rough flow chart to focus thinking.

3 Prioritize:
   a) Set boundaries to the investigation.
   b) Make use of ranking, Pareto, matrix analysis, etc., as appropriate.
   c) Review and gain agreement in the team of what is 'do-able'.

4 Define the task:
   a) Produce a written description of the process or problem area that can be confirmed with the team's sponsor.
   b) Confirm agreement in the team.
   c) This step may generate further questions for clarification by the sponsor of the process.

5 Agree success criteria:
   a) List possible success criteria. How will the team know when it has succeeded?
   b) Choose and agree success criteria in the team.
   c) Discuss and agree time scales for the project.
   d) Agree with 'sponsor'.
   e) Document the task definition, success criteria and time scale for the complete project.

Review

This stage is concerned with finding out what information is already available, gathering it together, structuring it, identifying what further information might be needed, and agreeing in the team *what* is needed, *how* it is going to be obtained, and *who* is going to get it.

1 Gather existing information:
   a) Locate sources – verbal inputs, existing files, charts, records, etc.
   b) Go and collect, ask, investigate.

2 Structure information:
   a) Information may be available but not in the right format, e.g. number of unsterile packs known, but machine number, product type and time of production mixed up.

3 Define gaps:
   a) Is enough information available?
   b) What further information is needed?

c) What equipment is affected?
d) Is the product/service from one plant or area?
e) How is the product/service at fault?

If the answer to any of the questions is 'Do not know' then:

4 Plan further data collection:
   a) Use any data already being collected.
   b) Draw up checksheet(s).
   c) Agree data collection tasks in the team – *who, what, how, when.*
   d) Seek to involve others where appropriate. Who actually has the information? Who really understands the process?
   e) This is a good opportunity to start to 'extend the team' and involve others in preparation for the *execute* stage later on.

Investigate

This stage is concerned with analysing all the data, considering all possible improvements, and prioritizing these to come up with one or more solutions to the problem, or improvements to the process, which can be verified as being the answer which meets the success criteria.

1 Implement data collection action plan:
   a) Check at an early stage that the plan is satisfying the requirements.

2 Analyse data:
   a) What picture is the data painting?
   b) What conclusions can be drawn?
   c) Use all appropriate tools to give a clearer picture of the process.

3 Generate potential improvements:
   a) Brainstorm improvements.
   b) Discuss all possible solutions.
   c) Write down all suggestions (have there been any from outside the team?).

4 Agree proposed improvements:
   a) Prioritize possible proposals.
   b) Decide what is achievable in what time scales.
   c) Work out how to test proposed solution(s) or improvement(s).
   d) Design check sheets to collect all necessary data.
   e) Build a checking/verifying plan of action.

Verify

This stage is concerned with testing the plans and proposals to make sure that they work before any commitments to major process changes. This may involve a relatively short discussion round a table in a meeting or lengthy pilot trials in a laboratory, office or even a main operations area or production plant.

1 Implement action plan:
   a) Carry out the agreed tests on the proposals.

2 Collect data:
   a) Consider the use of questionnaires if appropriate
   b) Make sure the check sheets are accumulating the data properly.

3 Analyse data.

4 Verify success criteria are met:
   a) Compare performance of new or changed process with success criteria from define stage.
   b) If success criteria are not met return to appropriate stage in drive model (usually the investigate) stage.
   c) Continue until the success criteria have been meet. For difficult problems, it may be necessary to go a number of times round this loop.

Execute

The stage is concerned with selling the solution or process improvements to others (e.g. the process owner) who may not have been involved in the investigation but whose commitment is vital to ensure success. Part of this stage may well be the need to address the existing documented management system.

1 Develop implementation plan to gain commitment:
   a) Is there commitment from others? Consider all possible impacts.
   b) Actions?
   c) Timing?
   d) Selling required?
   e) Training required for new or modified process?

2 Review appropriate system paperwork/documentation:
   a) Who should do this? The team? The activity/process owner?
   b) What are the implications for other systems?
   c) What controlled documents are affected?

3 Gain agreement to all facets of the execution plan from the process owner.

4 Implement the plan.

5 Monitor success:
   a) Extent of original team involvement? Initially perhaps and then at intervals?
   b) 'Delegate' to process owner/ department involved? At what stage?

	Define	Review	Investigate	Verify	Execute
Brainstorming / affinity diagram	●	●	●		
Cause and effect diagrams	●				
Pareto	●	●		●	●
Matrix analysis	●	●	●		
Checksheets		●	●	●	●
Flowcharts		●	●		●
Force-field		●			●
Scatter diagram		●		●	
Histograms		●			
Charts		●	●	●	●
Project bar chart				●	●

Figure 13.2   Likely tools in the drive model

6 Balance between team taking responsibility for meeting its agreed project success criteria and ownership within the organization of processes and continuous improvement.

The basic improvement tools (see Chapter 9) most likely to be used at each of the drive stages are shown in Figure 13.2. The position of the **drive** system in the breakdown of the most critical processes is represented in Figure 13.3.

An example of the drive model used in practice

The example below shows how the drive model for a particular project worked out in practice. The sort of responses made by the team are given in quotes thus: Stated task: 'We lose orders because our response time in making quotations is too long. We must significantly reduce our response time.'

1 Define stage
a) Interrogate the task:

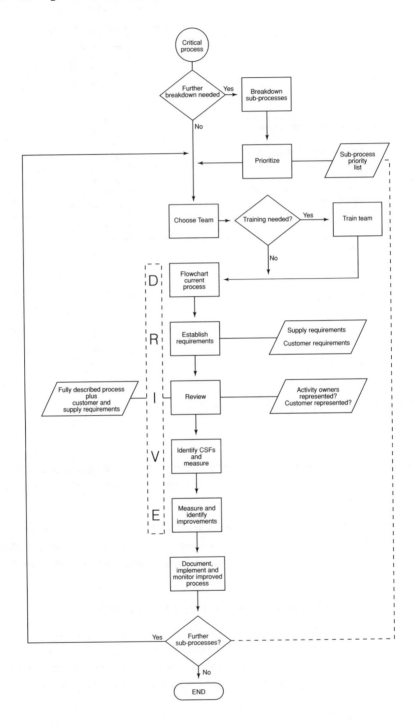

Figure 13.3 Critical processes and the drive model

1) Can we accept the problem as stated? 'Yes, this is generally known to be a problem area.'
2) Does this apply to all customers, or specific product lines? 'All customers.'
3) Does anyone measure response time at the moment? 'There was a one-off assessment a long time ago but it is not routinely measured.'
4) Have we the right expertise in our team to tackle it? 'Not really sure until we understand the problem.'
5) Can we succeed with this problem? 'Yes, if we don't let it get too big.'

b) Understand the process:
1) What processes 'contain' this problem? 'Customer visits by sales reps; telephone enquiries system; telex/fax enquiries; development department (technical vetting and provision of samples); pricing department'.
2) What is wrong at present? 'No real liaison between departments; laboratories have other priorities; visit reports not explicit.'

c) Prioritize:
1) What should be the boundaries of our investigation? 'Enquiries arising from direct sales visits to customers compose about 70 per cent of all enquiries. We will restrict our project to this area initially.'
2) Is it do-able? 'Yes, but now we realize we need someone from the development department on the team.'

d) Define the task.

e) Agree success criteria.

In response to d) and e) the team finally documented:

Customer quotations project
Our task is to investigate and reduce delays in the handling of those customer enquiries which arise from direct visits by our sales force. The project will be conducted in three phases:

1 a) To establish the average time (in days) between the salesman's visit and receipt by the customer of our quotation.
   b) To agree a target reduction in the average response time to enquiries.

2 To make recommendations which will enable the target reduction in response time to be achieved.

3 To implement the recommendations and monitor response times on a sample basis to demonstrate that the desired reduction has been achieved

Milestones for the completion of each phase, measured from the formal go-ahead date for these proposals, are:

Phase 1 – 1 month
Phase 2 – 4 months
Phase 3 – 8 months

2 Review stage

The team was not able to locate the original study report, but did discover a memo which gave the following summary:

• Average time to process a quotation request: 17 days

• Average time of quotations judged 'too late': 20 days

• Percentage of quotations 'too late': 30 %

From this, they concluded that the average time would have to be reduced to about 9 days, so that only 1 per cent of quotations would be 'too late', i.e. exceed 20 days.

3 Investigate stage

The team constructed flow charts of the various stages of the process. Major 'grey areas' occurred in dealing with quotations for new(er) products or new customers, where more technical vetting by the laboratory was required because: a) customer requirements were not clear, b) salesmen were not authorized to offer new specifications without checking with the technical department. It was in these areas that the sales visit reports were not sufficiently specific.

The flow chart of the process changed to include a path giving early warning to the technical department of requirements from 'new business areas' by identifying specifically named customers and product types. Quotations in these areas were treated with priority.

A check sheet to measure quotation turn-round times was designed.

4 Verify stage

• The modified process was implemented.

• The check sheet was used to gather data.

• A 'c chart' was used to monitor the average turn-round time in days for each week's orders, with the new procedure introduced at week 10. Action and warning lines for the chart were based on a target average of 9 days. The ensuing chart is shown in Figure 13.4.

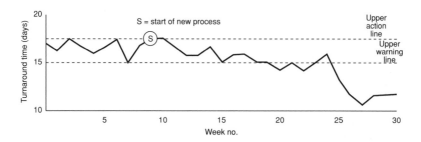

Figure 13.4   Charting the effects of improvement through **drive**

5 Execute stage

- The above data was presented to a meeting of the sales and technical departments.

- The change procedures were agreed, documented and circulated, including the list of current customers and products for special vetting.

- A procedure was documented to call for all quotations for new customers and products to be added to the special list and retained until familiarity enabled them to become 'standard'. They were then removed from the list for special attention.

- Continued monitoring using Figure 13.4 showed the average turn-round time reduce eventually to 10 days. Only 2 per cent of quotations were then taking greater than 20 days.

Steps in the introduction of continuous improvement teams

The idea of introducing continuous improvement teams often makes its way into an organization through the awareness of successful results in other organizations or companies. There is no fixed methodology for starting a teamwork programme, but there are certain key points which must be considered:

1   The concept should be presented to (or come from) management and supervision and their commitment and support enlisted. It should be possible at this stage to engage the interest and support of potential team leaders.

2   Projects should be started slowly and on a small scale. Ideally a pilot scheme, involving the most enthusiastic candidates and areas, should be

launched. Early teething troubles, doubts and worries may then be identified and resolved.

3   Selected or volunteer team leaders must be trained in all aspects of group
leadership, and the appropriate techniques, and they should be subsequently involved in the training of the team members in the techniques
required in effective problem solving. Systematic techniques should be
introduced, particularly brainstorming, cause and effect analysis, Pareto
analysis and charting. These concepts lay the groundwork for analysing
problems in a systematic fashion, and show that the majority of the problems are concentrated into a few areas.

4   Once the causes have been determined, a solution can be proposed. This
solution may involve any of the components of the process: equipment,
procedures, training, input requirements or output requirements. The proposed solution should be tested by the team, particularly if it involves procedures.

5   If the test of a solution proves successful, full-scale implementation can
then be carried out. In the case of procedures, full documentation of the
solution and management approval should be obtained. The procedure
can then be communicated to all personnel involved. Full-scale changes in
equipment and other processes should occur in the same manner. The
team should monitor implementation of the solution, plotting the appropriate data until the criteria for solution are met.

With the initial problems declared solved, the team may then tackle
another problem, and another, or be disbanded and new teams formed.
The record of successful solutions will motivate other teams within the
organization and ideas should spread. As the number of continuous
improvement teams in a company grows, new opportunities arise for
stimulating interest. Some large companies organize in-house conferences of
their improvement teams, providing the opportunity for the publication of
results and for recognition. Experience has shown that very significant improvements in areas such as energy reduction, productivity, and
cost-effectiveness may be achieved using the continuous improvement team
approach.

One of the problems of the team approach to improvement is that sometimes the teams are organized because it is the fashionable thing to do. They
either exist on paper only, or the meetings are social gatherings where nothing
is learned, no projects are initiated, and people do not grow. Another common
problem is that the teams attempt to solve process problems without first
obtaining knowledge of the necessary techniques: enthusiasm outruns ability.

Teams have enormous potential for helping to improve an organization's processes but for them to be successful they must follow a disciplined approach using proven techniques.

The team approach to continuous improvement works by tapping the skills and initiative of all personnel involved in a process. This may mean a change in culture which must be supported by management through their own activities and behaviour.

Statistical process control

The responsibility for improvement of any process must in the end lie with the operators of that process. To fulfil this responsibility, however, people must be provided with the tools necessary to:

- know whether the process is capable of meeting the requirements,

- know whether the process is meeting the requirements at any point in time,

- make correct adjustments to the process or its inputs when it is not meeting the requirements.

The techniques of statistical process control (SPC) will greatly assist in these stages.

All processes can be monitored and brought 'under control' by gathering and using data. This refers to measurement of the performance of the process and the feedback required for corrective action, where necessary. SPC methods, backed by management commitment and good organization, provide objective means of **controlling** quality in any transformation process, whether used in the manufacture of artefacts, the provision of services, or the transfer of information.

SPC is not only a tool kit, it is a strategy for reducing variability, the cause of many problems: variation in products, in times of deliveries, in ways of doing things, in materials, in people's attitudes, in equipment and its use, in maintenance practices, in everything. Control by itself is not sufficient. Total organizational excellence requires that the process should be improved continually by reducing its variability. This is brought about by studying all aspects of the process using the basic question: 'Could we do this job more consistently and on target?' the answering of which drives the search for improvements. This significant feature of SPC means that it is not constrained to measuring conformance, and that it is intended to lead to action on processes which are operating within the 'specification' to minimize variability.

Process control is essential and SPC should form a part of the overall organizational strategy. Incapable and inconsistent processes render the best design impotent and make supplier quality assurance irrelevant. Whatever process is being operated, it should be reliable and consistent. SPC can be used to achieve this objective.

In the application of SPC there is often an emphasis on techniques rather than on the implied wider managerial strategies. It is worth repeating that SPC is not only about plotting charts on the walls of a plant or office, it must become part of the company-wide adoption of continuous improvement. Changing an organization's environment into one in which SPC can operate properly may take several years rather than months. For many companies SPC will bring a new approach, a new 'philosophy', but the importance of the statistical techniques should not be underestimated. Simple presentation of data using diagrams, graphs, and charts can become the means of communication concerning the state of control of processes. It is on this understanding that improvements will be based.

The SPC system*

A systematic study of any process through answering the questions:

- Are we capable of doing the job correctly?

- Do we continue to do the job correctly?

- Have we done the job correctly?

- Could we do the job more consistently and on target?

provides knowledge of the **process capability** and the sources of non-conforming outputs. This information can then be fed back quickly to marketing, design and the 'technology' functions. Knowledge of the current state of a process also enables a more balanced judgement of equipment, both with regard to the tasks within its capability and its rational utilization.

Statistical process control procedures exist because there is variation in the characteristics of all material, articles, services and people. The inherent variability in every transformation process causes the output from it to vary over a period of time. If this variability is considerable, it is impossible to predict the value of a characteristic of any single item or at any point in time. Using statistical methods, however, it is possible to take meagre knowledge of the output and turn it into meaningful statements which may then be used to

---

*See John Oakland (1999) *Statistical Process Control – a really practical guide*, 4th edition, Butterworth-Heinemann, Oxford.

describe the process itself. Hence, statistically based process control procedures are designed to divert attention from individual pieces of data and focus it on the process as a whole. SPC techniques may be used to measure and control the degree of variation of any purchased materials, services, processes and products, and to compare this, if required, to previously agreed specifications. In essence, SPC techniques select a representative, simple, random sample from the 'population', which can be an input to or an output from a process. From an analysis of the sample it is possible to make decisions regarding the current performance of the process.

The management system and SPC in improving processes

The impact of an efficient management system and SPC together is that of gradually reducing process variability to achieve continuous or never-ending improvement. The requirement to set down defined procedures for all aspects of an organization's operations, and to stick to them, will reduce the variations introduced by the numerous different ways often employed of doing things. Go into any factory without a well defined and managed quality system in operation and ask to see the operators' 'black book' of plant operation and settings. Of course, each shift has a different black book, each with slightly different settings and ways of operating the process. Is it any different in office work or for sales people in the field? Do not be fooled by the perceived simplicity of a process into believing that there is only one way of operating it. There are an infinite variety of ways of carrying out the simplest of tasks – the author recalls seeing various course participants finding fourteen different methods for converting A4 size paper into A5 size (half A4) in a simulation of a production task. The ingenuity of human beings needs to be controlled if these causes of variation are not to multiply together to render processes completely incapable of consistency or repeatability.

The role of the management system here is to define and control processes, procedures and methods. Continual system audit and review will ensure that procedures are either followed or corrected, thus eliminating assignable or special causes of variation in materials, methods, equipment, information, etc., to ensure a 'could we do this job with more consistency?' approach.

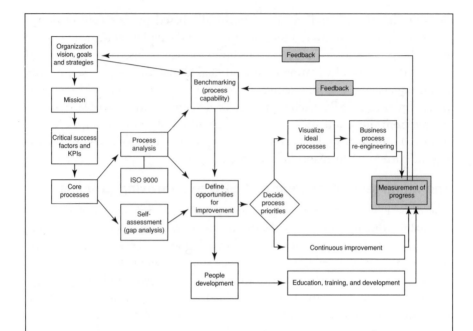

Performance measurement and feedback _____

Chapter 14

Performance measurement and feedback _____

Key points

Traditional performance measures based on cost accounting information provide little to support organizational excellence, because they do not map process performance and improvements seen by the customer. This chapter shows how performance measurement is important in identifying opportunities, and comparing performance internally and externally. A performance measurement framework will be recommended at the organizational, process, and individual levels, plus a review system.

The Deming cycle of plan, do, check, act (PDCA) is a useful design aid for measurement systems. It will be stressed that for performance measurement, the strategic objectives must be converted into critical success factors (CSFs), and appropriate key performance indicators (KPIs) developed.

It is often better to start with the simple measures and improve them, and the value of any measure must be compared with the cost of producing it.

The separation of process management and process performance is important and all critical parts of the process should be measured. Process owners should be involved in defining the performance measures which must reflect customer requirements.

The chapter also explains that performance review techniques, such as quality costing and self-assessment, are useful to identify improvement opportunities and motivate performance improvement. There is some discussion of the nature of the feedback to the benchmarking and strategic planning processes and how these may determine the course towards measured total organizational excellence.

Performance measurement and the improvement cycle

Traditionally, performance measures and indicators have been derived only from cost-accounting information, often based on outdated and arbitrary principles. These provide little motivation to support attempts to improve performance and, in some cases, actually inhibit continuous improvement because they are unable to map process performance. In the organization that is to succeed over the long term, performance must begin to be measured by the improvements seen by the customer. In the cycle of never-ending improvement, measurement plays an important role in:

- tracking progress against organizational goals

- identifying opportunities for improvement

- comparing performance against internal standards

- comparing performance against external standards.

The author and his colleagues have seen many examples of so-called performance measurement systems that frustrated improvement efforts. Various problems include systems that:

1 Produce irrelevant or misleading information.

2 Track performance in single, isolated dimensions.

3 Generate financial measures too late, e.g. quarterly, for mid-course corrections or remedial action.

4 Do not take account of the customer perspective, both internal and external.

5 Distort management's understanding of how effective the organization has been in implementing its strategy.

6 Promote behaviour which undermines the achievement of the strategic objectives.

Typical harmful summary measures of local performance are purchase price, machine or plant efficiencies, direct labour costs, and ratios of direct to indirect labour. These are incompatible with performance improvement measures such as process and throughput times, delivery performance, inventory reductions, and increases in flexibility, which are first and foremost *non-financial*. Financial summaries provide valuable information of course, but they should not be used for control. Effective decision-making requires direct physical measures for operational feedback and improvement.

One example of a 'measure' with these shortcomings is return on investment (ROI). ROI can be computed only after profits have been totalled for a given period. It was designed therefore as a single-period, long-term measure, but it is often used as a short-term one. Perhaps this is because most executive bonus 'packages' in the West are based on short-term measures. ROI tells us what happened, not what is happening or what will happen, and, for complex and detailed projects, ROI is inaccurate and irrelevant.

Many managers have a poor or incomplete understanding of their processes and products or services, and, looking for an alternative stimulus, become interested in financial indicators. The use of ROI, for example, for evaluating strategic requirements and performance can lead to a discriminatory allocation of resources. In many ways the financial indicators used in a large number of businesses have remained static, while the environment in which they operate has changed dramatically.

Traditionally, the measures used have not been linked to the processes where the value-adding activities take place. What has been missing is a performance measurement framework that provides feedback to people in all areas of business operations and stresses the need to fulfil customer needs.

The critical elements of a good performance measurement framework are:

- Leadership and commitment.
- Full employee involvement.
- Good planning.
- Sound implementation strategy.
- Measurement and evaluation.
- Control and improvement.
- Achieving and maintaining standards of excellence.

The Deming cycle of continuous improvement – plan, do, check, act – clearly requires measurement to drive it, and yet it is a useful design aid for the measurement system itself:

Plan: establish performance objectives and standards.

Do: measure actual performance.

Check: compare actual performance with the objectives and standards – determine the gaps.

Act: take the necessary actions to close the gaps and make the necessary improvements.

In this chapter a performance measurement framework is proposed, based on the strategic planning and process management models outlined in Chapters 3 and 5.

The framework has four elements related to: strategy development and goal deployment, process management, individual performance management, and review (Figure 14.1). This reflects an amalgamation of the approaches used by a range of organizations in performance measurement.

As we have seen in earlier chapters, the key to strategic planning and goal deployment is the identification of a set of critical success factors (CSFs) and associated key performance indicators (KPIs). These factors should be derived from the organization's mission, and represent a balanced mix of stakeholders. Action plans over both the short- and long-term should be developed, and responsibility clearly assigned for performance. The strategic goals of the organization should then be clearly communicated to all individuals, and translated into measures of performance at the process/functional level.

The key to successful performance measurement at the process level is the identification and translation of customer requirements and strategic objectives into an integrated set of process performance measures. The documentation and management of processes has been found to be vital in this translation process. Even when a functional organization is retained, it is nec-

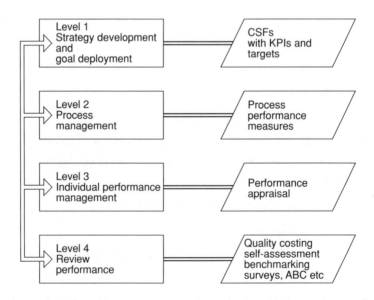

Figure 14.1 Performance measurement framework

essary to treat the measurement of performance between departments as the measurement of customer–supplier performance.

Performance measurement at the individual level usually relies on performance appraisal, i.e. formal planned performance reviews, and performance management, namely day-to-day management of individuals. A major drawback with some performance appraisal systems, of course, is the lack of their integration with other aspects of performance measurement.

Performance review techniques are used by many world class organizations to identify improvement opportunities, and to motivate performance improvement. These companies typically use a wide range of such techniques and are innovative in performance measurement in their drive for continuous improvement.

The links between performance measurement at the four levels of the framework are based on the need for measurement to be part of a systematic process of continuous improvement, rather than for 'control'. The framework provides for the development and use of measurement, rather than prescriptive lists of measures that should be used. It is, therefore, applicable in all types of organization.

The elements of the performance measurement are distinct from the budgetary control process, and also from the informal control systems used within organizations. Having said that, performance measurement should not be treated as a separate isolated system. Instead measurement is documented as and when it is used at the organizational, process and individual levels. In this way it can facilitate the alignment of the goals of all individuals, teams, departments and processes with the strategic aims of the organization and incorporate the voice of the stakeholders in all planning and management activities.

A number of factors have been found to be critical to the success of performance measurement systems. These factors include the level of top management support for non-financial performance measures, the identification of the vital few measures, the involvement of all individuals in the development of performance measurement, the clear communication of strategic objectives, the inclusion of customers and suppliers in the measurement process, and the identification of the key drivers of performance. These factors will need to be taken into account by managers wishing to develop a new performance measurement system, or refine an existing one.

The performance measurement framework

In most organizations there are no separate performance measurement systems. Instead, performance measurement forms part of wider organizational management processes. Although elements of measurement can be identified at many different points within organizations, measurement itself usually forms the 'check' stage of the continuous improvement PDCA cycle. This is important since measurement data that is collected but not acted upon in some way is clearly a waste of resources.

The four elements of the framework in Figure 14.1 are:

Level 1 Strategy development and goal deployment leading to mission/vision, critical success factors and **key performance indicators** (KPIs).

Level 2 Process management and **process performance measurement** (including input, in-process and output measures, management of internal and external customer–supplier relationships and the use of management control systems).

Level 3 Individual performance management and **performance appraisal**.

Level 4 Review performance (including internal and external **benchmarking, self-assessment** against quality award criteria and **quality costing**).

Level 1 – Strategy development and goal deployment

The first level of the performance measurement framework is the development of organizational strategy, and the consequent deployment of goals throughout the organization. Steps in the strategy development and goal deployment measurement process are (see also Chapter 3):

1 Develop a mission statement based on recognizing the needs of all organizational stakeholders, customers, employees, shareholders and society.

2 Based on the mission statement, identify those factors critical to the success of the organization achieving its stated mission. Again CSFs should represent all the stakeholder groups, customers, employees, shareholders and society.

3 Define performance measures for each CSF, i.e. key performance indicators (KPIs). There may be one or several KPIs for each CSF.
Definition of KPIs should include:
   a) title of KPI.
   b) data used in calculation of KPI.
   c) method of calculation of KPI.
   d) sources of data used in calculation.
   e) proposed measurement frequency.
   f) responsibility for the measurement process.

4 Set targets for each KPI. If KPIs are new, targets should be based on customer requirements, competitor performance or known organizational criteria. If no such data exists, a target should be set based on best guess criteria. If the latter is used, the target should be updated as soon as enough data is collected to be able to do so.

5 Assign responsibility at the organizational level for achievement of desired performance against KPI targets. Responsibility should rest with directors and very senior managers.

6 Develop plans to achieve the target performance. This includes both action plans for one year, and longer-term strategic plans.

7 Deploy mission, CSFs, KPIs, targets, responsibilities and plans to the core business processes. This includes the communication of goals, objectives, plans, and the assignment of responsibility to appropriate individuals.

8 Manage organizational processes (see Level 2 of the framework).

9 Measure performance against organizational KPIs, and compare to target performance.

10 Based on this comparison, identify areas with high leverage for improvement, and update action plans.

11 Communicate performance and proposed actions throughout the organization.

12 At the end of the planning cycle compare organizational capability to target against all KPIs, and begin again at Step 2 above.

13 Reward and recognize superior organizational performance.

Strategy development and goal deployment are clearly the responsibilities of senior management within the organization, although there should be as much input to the process as possible by employees to achieve 'buy-in' to the process.

The system outlined above is similar to the policy deployment approach known as Hoshin Kanri, developed in Japan and adapted in the West.

Key performance indicators (KPIs)
The derivation of KPIs may follow the 'balanced scorecard' model, proposed by Kaplan, which divides measures into financial, customer, internal business and innovation and learning perspectives (Figure 14.2).

A balanced scorecard derived from the business excellence model described

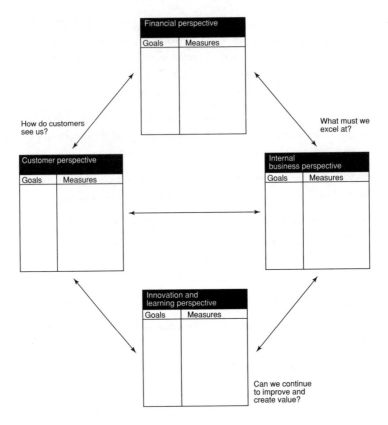

Figure 14.2 The balanced scorecard linking performance measures

in Chapter 7 would include financial and non-financial results, customer satisfaction (measured via the use of customer satisfaction surveys and other measures, including quality and delivery), employee factors (employee development and satisfaction), and societal factors (including community perceptions and environmental performance).

Financial performance for external reporting purposes may be seen as a result of performance across the other KPIs, the non-financial KPIs assumed to be the leading indicators of performance. The only aspect of financial performance that is cascaded throughout the organization is the budgetary process, which acts as a constraint rather than a performance improvement measure.

In summary then, organizational KPIs should be derived from the balancing of internal capabilities against the requirements of identified stakeholder groups. This has implications for both the choice of KPIs and the setting of appropriate targets. There is a need to develop appropriate action plans and clearly define responsibility for meeting targets if the KPIs and targets are to be taken seriously.

Level 2 – Process management and measurement

The second level of the performance measurement framework is process management and measurement, the steps of which are:

1 If not already completed, identify and map processes. This information should include identification of:
   a) Process customers and suppliers (internal and external)
   b) customer requirements (internal and external)
   c) core and non-core activities
   d) measurement points and feedback loops.

2 Translate organizational goals, action plans and customer requirements into process performance measures (input, in-process and output). This includes definition of measures, data collection procedures, and measurement frequency.

3 Define appropriate performance targets, based on known process capability, competitor performance and customer requirements.

4 Assign responsibility for achieving performance targets.

5 Develop plans towards achievement of process performance targets.

6 Deploy measures, targets, plans and responsibility to all sub-processes.

7 Operate processes.

8 Measure process performance and compare to target performance.

9 Use performance information to:
   a) implement continuous improvement activities.
   b) identify areas for improvement.
   c) update action plans.
   d) update performance targets.
   e) redesign processes, where appropriate.
   f) manage the performance of teams and individuals (performance management and appraisal) and external suppliers.
   g) provide leading indicators and explain performance against organizational KPIs.

10 At the end of each planning cycle compare process capability to customer requirements against all measures, and begin again at Step 2.

11 Reward and recognize superior process performance, including sub-processes, and teams.

The same approach should be deployed to sub-processes and to the activity and task levels.

The above process should be managed by the process owner, with inputs wherever possible from the owners of sub-processes. The process outlined should be used whether an organization is organized and managed on a process or functional departmental basis. If functionally organized, the key task is to identify the customer–supplier relationships between functions, and for functions to see themselves as part of a customer–supplier chain.

Performance measures

Performance measures used at the process level differ widely between different organizations. Some organizations measure process performance using a balanced scorecard approach, whilst others monitor performance across different dimensions according to the process. Whichever method is used, measurements should be identified as input (supplier), in-process, and output (or results-customers).

It is usually at the process level that the greatest differences can be observed between the measurement used in manufacturing and services organizations. However, all organizations should measure quality, delivery, customer service/satisfaction, and cost.

Depending on the process, measurement frequency varies from daily, for example in the measurement of delivery performance, to annual, for example in the measurement of employee satisfaction, which has implications for the PDCA cycle time of the particular process(es). Measurement frequency at the process level may, of course, be affected by the use of information technology. Cross-functional process performance measurement is a vital component in the removal of 'functional silos', and the consequent potential for sub-optimization and failure to take account of customer requirements. The success of performance measurement at the process level is dependent on the degree of management of processes and on the clarity of the deployment of strategic organizational objectives.

Measuring and managing the whats and the hows

Busy senior management teams find it useful to distil as many things as possible down to one piece of paper or one spreadsheet. The use of KPIs, with targets, as measures for CSFs, and the use of performance measures for processes may be combined into one matrix which is used by the senior management team to 'run the business'.

Figure 14.3 is an example of such a matrix which is used to show all the useful information and data needed:

Core processes					Manage people	Develop products	Develop new business	Manage our accounts	Manage financials	Manage int. systems	Conduct research
CSFs: We must have	Measures	Year targets	Target CSF owner								
Satisfactory financial and non-financial performance	Sales volume. Profit. Costs versus plan. Shareholder return Associate/employee utilisation figures	Turnover £2m. Profit £200k. Return for shareholders. Days/ month per person			X		X	X	X	X	
A growing base of satisfied customers	Sales/customer Complaints/recommendations Customer satisfaction	>£200k = 1 client. £100k-£200k =5 clients £50k-£100k= 6 clients <£50k=12 clients			X	X	X	X		X	X
A sufficient number of committed and competent people	No. of employed staff/associates Gaps in competency matrix. Appraisal results Perceptions of associates and staff	15 employed staff 10 associates including 6 new by end of year			X				X	X	
Research projects properly completed and published	Proportion completed on time, in budget with customers satisfied. Number of publications per project	3 completed on time, in budget with satisfied customers			X			X			X
** = Priority for improvement					**		**	**			
Process owner											
Process performance											
Measures and targets											

Figure 14.3 CSF/core process reporting matrix

- the CSFs and their owners   the *whats*
- the KPIs and their targets
- the core business processes and their sponsors the *hows*
- the process performance measures

It also shows the impacts of the core processes on the CSFs. This is used in conjunction with a 'business management calendar', which shows when to report/monitor performance, to identify process areas for improvement. This slick process offers senior teams a way of:

- gaining clarity about what is important and how it is measured;
- remaining focused on what is important and what the performance is;
- knowing where to look if problems occur.

Level 3 – Individual performance and appraisal management

The third level of the performance measurement framework is the management of individuals. Performance appraisal and management are usually the responsibility of the direct managers of individuals whose performance is to be appraised. At all stages in the process, the individuals concerned must be included to ensure 'buy in'.

Steps in performance and management appraisal are:

1 If not already completed, identify and document job description based on process requirements and personal characteristics. This information should include identification of:
   a) activities to be undertaken in performing the job;
   b) requirements of the individual with respect to the identified activities, in terms of experience, skills and training;
   c) requirements for development of the individual, in terms of personal training and development;

2 Translate process goals and action plans, and personal training and development requirements into personal performance measures.

3 Define appropriate performance targets based on known capability and desired characteristics (or desired characteristics alone if there is no prior knowledge of capability).

4 Develop plans towards achievement of personal performance targets.

5 Document 1 to 4 using appropriate forms, which should include space for the results of performance appraisal.

6 Manage performance. This includes:
   a) planning tasks on a daily/weekly basis;
   b) managing performance of the tasks;
   c) monitoring performance against task objectives using both quantitative (process) and qualitative information on a daily and/or weekly basis;
   d) giving feedback to individuals of their performance in carrying out tasks;
   e) giving recognition to individuals for superior performance.

7 Formally appraise performance against range of measures developed, and compare to target performance.

8 Use comparison with target to:
   a) identify areas for improvement;
   b) update action plans;
   c) update performance targets;
   d) redesign jobs, where appropriate. This impacts Step 1 of the process.

9 Update documentation.

10 After a suitable period, ideally more than once a year, compare capability to job requirements and begin again at Step 2.

11 Reward and recognize superior performance.

The above activities should be undertaken by the individual whose performance is being managed, together with their immediate superior.

The major differences in approaches in the management of individuals lies in the reward of effort as well as achievement and the consequently different measures used, and in the use of information in continuous improvement required to reward and recognize performance, including teamwork. Unlike management by objectives (MBO), where the focus is on measurement of results – which are often beyond the control of the individual whose performance is appraised – good performance management systems attempt to measure a combination of process/task performance (effort and achievement) and personal development.

The frequency of formal performance appraisal is generally defined by the frequency of the appraisal process usually with a minimum frequency of six months. Between the formal performance appraisal reviews, most organizations rely on the use of other performance management techniques to manage individuals. Measures of team performance, or of participation in teams, should be included in the appraisal systems where possible, to improve team performance. In many organizations, the performance appraisal system is probably the least successfully implemented element of the framework. Appraisal systems are often designed to motivate individuals to achieve process and personal development objectives, but not to perform in teams. One of the limitations of appraisal processes is the frequency of measurement, which could be increased, but few organizations would consider doing so.

Level 4 – Performance review

The fourth level of the performance measurement framework is the use of performance review techniques. Steps in review are as follows:

1   Identify the need for review, which may come from:
    a)   poor performance at the organizational or process levels against KPIs;
    b)   identified superior performance of competitors;
    c)   customer inputs;
    d)   the desire to better direct improvement efforts;
    e)   the desire to concentrate attention on the need for performance improvement.

2   Identify method of performance review to be used. This involves determining whether the review should be carried out internally within the organization, or externally, and the method that should be carried out. Some techniques are mainly internal, e.g. self-assessment, quality costing; whilst others, e.g. benchmarking, involve obtaining information from sources external to the organization. The choice should depend on:
    a)   how the need for review was identified (see Step 1);
    b)   the aim of the review, e.g. if the aim is to improve performance relative to competitors, external benchmarking may be a better option than internally measuring the cost of quality;
    c)   the relative costs and expected benefits of each technique.

3   Carry out the review.

4   Feed results into the planning process at the organizational or process level.

5   Determine whether to repeat the exercise. If it is decided to repeat the exercise, the following points should be considered:
    a)   frequency of review;
    b)   at what levels to carry out future reviews, e.g. organization-wide or process-by-process;
    c)   decide whether the review technique should be incorporated into regular performance measurement processes, and if so how this will be managed.

Review methods often require the use of a level of resources greater than that normally associated with performance measurement, often due to the need to develop data collection procedures, train people in their use, and the cost of data collection itself. However, review techniques usually give a broader view of performance than most individual measures.

The use of review techniques is most successful when it is based on a clearly identified need, perhaps due to perceived poor performance against existing performance measures or against competitors, and the activity itself is clearly planned and the results used in performance improvement. This is often the difference between the success and failure of quality costing and benchmarking in particular. The use of most of the review techniques has been widely documented, but often without regard to their integration into the wider processes of measurement and management.

Review techniques
Techniques identified for review include:

1 Quality costing, using either prevention–appraisal–failure, or process costing methods.

2 Self-assessment against Baldrige, European Quality Award, or internally developed criteria.

3 Benchmarking, internal or external.

4 Customer satisfaction surveys.

5 Activity based costing (ABC)

Summarizing performance measurement and feedback to the total organizational excellence model

The budgetary control process – often the most clearly identifiable aspect of management control within organizations – is separate from the performance framework. The information in the framework is used in performance improvement, whereas budgetary control acts as a constraint within which performance is managed. It is vital that performance measurement is integrated into the overall management process, and that the data is used sensibly to manage the continuous improvement of performance.

Performance measurement should be treated as a resource consuming activity, so that decisions made regarding the number of measures to be collected, the frequency of data collection, and the criticality of the information are sensible. The frequency of the measurement cycle depends on a number of factors. The different levels go through different PDCA cycles. For example, a daily PDCA cycle in a manufacturing unit will be very different to the annual cycle at the organizational level. The important factor is that the data is used in a constructive, systematic manner, to generate actions which improve performance.

Feedback on performance should be made to the total organizational excellence framework and to two components in particular – benchmarking and strategic planning. Feedback on process performance capability, whether after a re-engineering effort or just following a continuous improvement regime, or people development activities, needs to go into the benchmarking effort to confirm that improvement against the benchmarks is actually taking place.

Overall organizational performance should inform the strategic planning process, for it is possible that re-visioning of the business as a whole is appropriate following assessment of where we are.

The total organizational excellence framework offers a comprehensive assembly of some of the major business 'buzz words' and fashionable programmes engaged in by many organizations during the 1980s and 1990s. Strategic planning, total quality management, process management, business excellence, self-assessment, benchmarking, business process re-engineering, continuous improvement and performance measurement have all had their high times and low times in some organizations. Putting them all together in a holistic view of the business can make the difference between a good organization and a world class one.

Bibliography

Adair, J. (1988) *Effective Leadership*, 2nd edition, Pan Books, London.

Adair, J. (1987) *Effective Teambuilding*, 2nd edition, Pan Books, London.

Adair, J. (1987) *Not Bosses but Leaders: How to lead the successful way*, Talbot Adair Press, Guildford.

Adair, J. (1988) *The Action-Centred Leader*, Industrial Society, London.

Born, G. (1994) *Process Management to Quality Improvement*, John Wiley, Chichester.

Braganza, A. and Myers, A. (1997) *Business Process Redesign – a view from the inside*, International Thomson Business Press, London.

Briggs Myers, I. (1987) *Introduction to Type: A description of the theory and applications of the Myers Briggs Type Indicator*, Consulting Psychologists Press, Palo Alto, CA.

Briggs Myers, I. and P. B. (1993) *Gifts Differing – Understanding Personality Type*, Consulting Psychologists Press, Palo Alto, CA.

British Quality Foundation (BQF) (2001) *The Model in Practice – using the EFQM Excellence Model to deliver continuous improvement*, London.

British Standard Institution (BSI) (2000) BS EN ISO 9000: 2000 – Quality Management Systems, London.

Brown, M.G. (1992) *Baldrige Award Winning Quality: How to Interpret the Malcolm Baldrige Award Criteria*, 2nd edition, ASQC, Milwaukee, WI.

Camp, R.C. (1995) *Business Process Benchmarking: finding and implementing best practices*, ASQC Quality Press, Milwaukee, WI.

Collins, J.C. and Porras, J.I. (1998) *Built to Last*, Random House, London.

Dale, B.G., Cooper, C. and Wilkinson, A. (1992) *Total Quality and Human Resources – A Guide to Continuous Improvement*, Blackwell, Oxford.

Dale, B.G. (ed) (1999) *Managing Quality*, 3rd edition, Philip Allan, Hemel Hempstead.

Deming, W.E. (1982) *Out of the Crisis*, MIT, Cambridge, Mass.

Deming, W.E. (1993) *The New Economics*, MIT, Cambridge, Mass.

Dimaxcescu, D. (1992) *The Seamless Enterprise – Making Cross Functional Management Work, Harper Business*, New York.

Dixon, J.R., Nanni, A. and Vollmann, T.E. (1990) *The New Performance Challenge – Measuring Operations for World Class Competition*, Business One, Irwin, Homewood, IL.

European Foundation for Quality Management (2001) *The EFQM Excellence Model – various publications*, EFQM, Brussels.

Federal Information Processing Standard (FIPS) (1993) Publications 183 and 184, National Institute of Standards and Technology (NIST), US.

Feigenbaum, A.V. (1991) *Total Quality Control*, 3rd edition, revised, McGraw-Hill, New York.

Francis, D. (1990) *Unblocking the Organisational Communication*, Gower, Aldershot.

Hall, R.W., Johnson, H.Y. and Turney, P.B.B. (1991) *Measuring Up – Charting Pathways to Manufacturing Excellence*, Business One, Irwin, Homewood, IL.

Hammer, M. and Champy, J. (1993) *Reengineering the Corporation – A Manifesto for Business Revolution*, Nicholas Brealey, London.

Harrington, H.J. (1995) *Total Improvement Management*, McGraw-Hill, New York.

Harrington, H.J. (1991) *Business Process Improvement*, McGraw-Hill, New York.

Hart, W.L. and Bogan, C.E. (1992) *The Baldrige: what it is, how it's won, how to use it to improve quality in your company*, McGraw-Hill, New York.

Hussey, D. (1998) *Strategic Management*, 4th edition, Butterworth-Heinemann, Oxford.

Hutchins, D. (1992) *Achieve Total Quality*, Director Books, Cambridge (UK).

Hutchins, D. (1990) *In Pursuit of Quality*, Pitman, London.

Ishikawa, K. (translated by D.J. Lu) (1985) *What is Total Quality Control? – The Japanese Way*, Prentice-Hall, Englewood Cliffs, NJ.

Jacobson, I. (1995) *The Object Advantage – Business Process Reengineering with object technology*, Addison-Wesley, Wokingham.

Johansson, H.J., McHugh, P., Pendlebury, A.J. and Wheeler, W.A. (1993) *Business Process Reengineering – Breakpoint strategies for market dominance*, John Wiley, Chichester.

Johnson, G. and Scholes, K. (2001) *Exploring Corporate Strategy*, (6th edition), Prentice Hall, London.

Joiner, B.L. (1994) *Fourth Generation Management – The New Business Consciousness*, McGraw-Hill, New York.

Juran, J.M. (1989) *Juran on Leadership for Quality: An Executive Handbook*, The Free Press (Macmillan), New York.

Kaplan, R.W. (ed) (1990) *Measures for Manufacturing Excellence*, Harvard Business School Press, Boston, Mass.

Katzenbach, J. R. and Smith, D. K. (1994) *The Wisdom of Teams*, McGraw-Hill, New York.

Kay, J. (1995) *Foundations of Corporate Success*, Oxford University Press, Oxford.

Kormanski, C. *A Situational Leadership Approach to Groups Using the Tuckman Model of Group Development*, The 1985 Annual Developing Human Resources Conference, University Associates, San Diego.

Kormanski, C. and Mozenter, A. *A New Model of Team Building: A Technology for Today and Tomorrow,* The 1987 Annual Developing Human Resources Conference, University Associates, San Diego.

Krebs Hirsh, S. (1992) *MBTI Team Building Program, Team Member's Guide,* Consulting Psychologists Press, Palo Alto, CA.

Krebs Hirsh, S., and Kummerow, J.M. (1987) *Introduction to Type in Organisational Settings,* Consulting Psychologists Press, Palo Alto, CA.

Larkin, T. J. and S. L. (1994) *Communicating Change,* McGraw-Hill, New York.

Macdonald, J. and Tanner, S. (1996) *Understanding Benchmarking in a Week,* Hodder and Stoughton.

McCaulley, M.H. (1975) How individual differences affect health care teams, *Health Team News,* **1**(8), pp. 1–4.

Merli, G. (1996) *Managing by Priority – Thinking strategically, acting effectively,* John Wiley, Chichester.

Mills Steeples, M. (1992) *The Corporate Guide to the Malcolm Baldrige National Quality Award,* ASQC, Milwaukee, WI.

Muhlemann, A.P., Oakland, J.S. and Lockyer, K.G. (1992) *Production and Operations Management,* 6th edition, Pitman, London.

Musselwhite, E. (1988) *Interpersonal Dimensions – Understanding Your FIRO-B Results,* Consulting Psychologists Press, Palo Alto, CA.

National Institute of Standards and Technology, USA (2001) *Baldrige National Quality Award – 2001 Criteria for Performance Excellence,* NIST, Gaithersburg.

Neave, H. (1990) *The Deming Dimension,* SPC Press, Knoxville.

Oakland, J.S. (2001) *Statistical Process Control: A Really Practical Guide,* 5th edition, Butterworth-Heinemann, Oxford.

Oakland, J.S. (1994) *Total Quality Management – The route to improving performance,* 2nd edition, Butterworth-Heinemann, Oxford.

Oakland, J.S. (2000) *Total Quality Management – Text with cases,* 2nd edition, Butterworth-Heinemann, Oxford.

Pitts, C. (1995) *Motivating your Organisation – Achieving business success through reward and recognition,* McGraw-Hill, Maidenhead.

Porter, L.J., Oakland, J.S. and Gadd, K.W. (1998) *Evaluating the Operation of the European Quality Award Model for Self-assessment,* CIMA Publishing, London.

Porter, L. and Tanner, S. (1996) *Assessing Business Excellence,* Butterworth-Heinemann, Oxford.

Ranjit Roy (1990) *A Primer on the Taguchi Method,* Van Nostrand Reinhold, New York.

Rummler, G.A. and Brache, A.P. (2000) *Improving Performance: how to manage the white space on the organisation chart,* 3rd edition, Jossey-Bass Publishing, San Francisco, CA.

Ryuji Fukuda (1990) *CEDAC – A Tool for Continuous Systematic Improvement,* Productivity Press, Cambridge, Mass.

Scholtes, P.R. (1990) *The Team Handbook*, Joiner Associates, Madison, NY (USA).

Schutz, W. (1958) *FIRO: A Three-Dimensional Theory of Interpersonal Behaviour*, Mill Valley WSA, CA.

Schutz, W. (1978) *FIRO Awareness Scales Manual*, Consulting Psychologists Press, Palo Alto, CA.

Schutz, W. (1994) *The Human Element – Productivity, Self-esteem and the Bottom Line*, Jossey-Bass, San Francisco, CA.

Sengi, P., Roberts, C., Ross, R.B., Smith, B.J. and Kleiner, A. (1994) *The Fifth Discipline Fieldbook – Strategies and Tools for Building a Learning Organisation*, Nicholas Brealey, London.

Spendolini, M.J. (1992) *The Benchmarking Book*, ASQC, Milwaukee, WI.

Stahl, M.J. (1995) *Management – Total Quality in a Global Environment*, Blackwell, Cambridge Mass.

Talley, D.J. (1991) *Total Quality Management: Performance and Cost Measures*, ASQC, Milwaukee, WI.

Townsend, P.L. and Gebhardt, J.E. (1992) *Quality in Action – 93 Lessons in Leadership, Participation and Measurement*, John Wiley, New York.

Tuckman, B.W. and Jensen, M.A. (1977) Stages of Small Group Development Revisited, *Group and Organisational Studies*, **2**(4), pp. 419–427.

Wellins, R.S., Byham, W.C. and Wilson, J.M. (1991) *Empowered Teams*, Jossey Bass, Oxford.

Whitley, R. (1991) *The Customer Driven Company*, Business Books, London.

White, A. (1996) *Continuous Quality Improvement*, Piatkus, London.

Zairi, M. (1996) *Benchmarking for Best Practice,* Butterworth-Heinemann, Oxford.

Zeithaml, V.A., Parasuraman, A. and Berry, L.L. (1990) *Delivering Quality Service: Balancing customer perceptions and expectations*, The Free Press (Macmillan), New York.

Index _____